T0258897

ON FUNCTIONS AND FUNCTIONAL EQUATIONS

Jaroslav Smítal

ON FUNCTIONS
AND FUNCTIONAL EQUATIONS

Taylor & Francis
Taylor & Francis Group
New York   London

Taylor & Francis is an imprint of the
Taylor & Francis Group, an informa business

British Library Cataloguing in Publication Data

Smítal, Jaroslav
On functions and functional equations.
1. Functional equations
I. Title   II. O funkciách a funkcionálnych rovniciach.
English
515.7   QA431

ISBN 0-85274-418-8

Library of Congress Cataloging-in-Publication Data

Smital, Jaroslav.
    On functions and functional equations.

    Bibliography: p. 150—152
    Includes index.
    1. Functional equations.   2. Iterative methods
(Mathematics.) 3. Functions of several real variables. I. Title.
QA431.S545   1988   515.7   87-21133

ISBN 0-85274-418-8

Translation: J. Dravecký

IOP Publishing Ltd
Techno House. Redcliffe Way. Bristol BS1 6NX, England
Published in co-edition with Alfa, Publishers of Technical
and Economical Literature, Bratislava

CONTENTS

PREFACE

Functional equations and iterations form a modern mathematical discipline, which has developed very rapidly in the last decade. The fertile soil on which this discipline has grown is the practice of such problems in physics, biology, and even in economics. However, mathematicians are also attracted to functional equations by their apparent simplicity and harmonic nature, which may conceal the possibility of obtaining important mathematical results. Some of these discoveries are discussed here, where the intention is to provide an elementary acquaintance with functional equations and iterations.

The introductory chapter summarizes the basic ideas necessary for the later parts of the book. Almost all of them are taught already to secondary school students. Readers well acquainted with continuity and derivatives of real functions in one (real) variable may omit this chapter. The central part of the book is focused on Chapter 3, which concerns iterations. Their applications are treated in Chapter 4. The next chapter (Chapter 5), on functional equations in one variable, also makes use of iterates. Functional equations in several variables are discussed, using simple tools, immediately after the introductory part, in Chapter 2, which is in fact an independent part of the book.

The book is written for readers who have taken secondary-school mathematics to a high level and are now continuing as undergraduates. A few sections and exercises are more exacting — these are marked with an asterisk and may be omitted in an introductory course.

I owe thanks to P Brunovský and P Kostyrko, the referees of the Slovak edition, for their valuable comments and advice that helped to improve the text in many places; to W Jarczyk, M Kuczma, A N Šarkovskii and Ľ Snoha for their constructive criticism of the Slovak edition and to J Dravecký, who translated the book into English.

Bratislava, January 1987

J Smítal

1. FUNCTIONS

1.1 ELEMENTARY PROPERTIES OF FUNCTIONS

Functional equations and their solutions cannot be discussed without understanding the notion of a function. Thus in this chapter we shall summarise precisely what we mean when using the terms function, continuity and derivative. Most of these things are covered by the secondary school mathematics curriculum and readers familiar with them may omit this introductory chapter and proceed to Chapter 2.

Let us begin with the basic concept.

Definition 1. Let A and B be sets. Then f is a function mapping A into B if exactly one $y = f(x) \in B$ is assigned to each $x \in A$.

The set A is then referred to as the domain of f and the set B is the range of f. We sometimes write $f: A \rightarrow B$, meaning that f maps A into B. If $f(x) = y$, then y is said to be the image of x and x is called a pre-image of y. In this book, we shall mostly deal with functions whose domain and range are subsets of the set R of all real numbers. They are termed real functions.

A function may be defined by a propositional form such as $y = 3x + 2x^2 - 1$ or $y = x \tan x$ in the explicit form, or by a rule like

$$\frac{x + y}{xy} = 1 \qquad \text{or} \qquad xy = 3x^2$$

1

in the implicit form. In the latter case, however, it is necessary to check whether the formula does define a function. For example, no function is defined by

$$x^2 + 2y^2 = 3.$$

(Why?)

Other ways of defining a function are

$$f(x) = \begin{cases} 3x & \text{for} \quad x < 0 \\ \sin \pi x & \text{for} \quad x \in [0, 1] \\ x^2 - 1 & \text{for} \quad x > 1 \end{cases} \qquad (1.1)$$

or

$$f(x) = \begin{cases} 0 & \text{if } x \text{ is an irrational number} \\ 1/q & \text{if } x \text{ is a rational number } p/q, \\ & \text{where } p \text{ and } q \text{ are coprime} \\ & \text{numbers and } q > 0. \end{cases} \qquad (1.2)$$

If the domain of a function defined by a propositional form is not given, then the maximal domain is understood. It is the largest set with the property that each of its elements can be 'substituted' in the form.

To every real function $f: A \to B$, we may assign a graph Γ_f. This is the set of all the points in the plane with coordinates $[t, f(t)]$, all the points of the domain A being taken as t. In this book, however, the graph of f will sometimes mean the set of all ordered pairs $[t, f(t)]$. Thus the graph of a function may be both a pictorial sketch and a set*. There are functions whose graph is difficult or even impossible to draw. The function (1.2) provides an example.

If $f: A \to B$ and A_1 is a subset of A, then $f(A_1)$ denotes the set of all those $y \in B$ that can be expressed in the form

---

* In school mathematics, a function is defined as a set of ordered pairs; for our purpose it is more convenient to consider a function as a rule — $f(t)$ changes with changing t.

$y = f(x)$ with $x \in A_1$. The set $f(A_1)$ may (and often will) also be written in the form

$$f(A_1) = \{f(x) \in B; \ x \in A_1\}$$

or just

$$f(A_1) = \{f(x); \ x \in A_1\}.$$

The set $f(A_1)$ is called the image of A_1. Similarly, if $B_1 \subset B$, then $f^{-1}(B_1)$ denotes the set $\{x \in A; f(x) \in B_1\}$ of all those elements of A which are mapped by f into B_1. The set $f^{-1}(B_1)$ is referred to as the pre-image of B_1.

If f is a function with domain A and range B and if $f(A) = B$, then f is said to be a mapping of A onto B (or a surjection).

In fact, every function is a surjection when a convenient range is considered. More precisely, if $f: A \to B$ is any function, it is a mapping of A onto $f(A) \subset B$. Take the function defined by equation (1.2) as an example. Considered as a mapping of R into R, where R is the set of all real numbers, f is not a surjection. But the same function maps R onto B, where B is the set containing only 0 and all the numbers $1/n$ with $n = 1, 2, \ldots$.

We say that a real function f defined on a set A is:

1. increasing on A if $f(x) < f(y)$ when $x, y \in A$, $x < y$;
2. decreasing on A if $f(x) > f(y)$ when $x, y \in A$, $x < y$;
3. non-increasing on A if $f(x) \geqslant f(y)$ when $x, y \in A$, $x < y$;
4. non-decreasing on A if $f(x) \leqslant f(y)$ when $x, y \in A$, $x < y$.

Functions of any of the types 1 to 4 are referred to as monotone functions, those of the types 1 and 2 are strictly monotone. Note that a function which is not increasing need not be non-increasing.

Example 1. The function $f(t) = t/(t + 1)$ is increasing on the interval $A_1 = (-\infty, -1)$, as well as on the interval $A_2 = (-1, \infty)$, but it is not increasing (not even monotone) on its maximal domain $A = A_1 \cup A_2$.

Example 2. The function g given by $g(t) = 0$ for $t \leqslant 1$ and $g(t) = t - 1$ for $t > 1$ is non-decreasing, but it is not increasing.

We also say that a real function $f: A \to B$ is:

1. upper bounded on A if there exists a constant (number) K such that $f(t) < K$ for all $t \in A$;
2. lower bounded on A if there is a number K with $f(t) > K$ for all $t \in A$;
3. bounded on A if it is both upper and lower bounded on A.

Thus a function f is bounded on a set A if there exist constants K_1 and K_2 such that

$$K_1 < f(t) < K_2$$

for all $t \in A$.

If $f: A \to B$ and $g: B \to C$ are two functions, then their composite $g \circ f: A \to C$ is defined by

$$(g \circ f)(t) = g(f(t)).$$

The image of t with respect to the composite function $g \circ f$ is found by first finding its f-image (that is, the image of t with respect to f) and then the g-image of $f(t)$, that is $g(f(t))$. For a composition of two functions, it is very important that the range of the first of them be a subset of the domain of the second function. However, we may compose functions even if this condition is not met, and this is a common practice. Then, of course, the domain of the composite becomes smaller. In the extreme case it may happen that the domain of the composite is the empty set $\emptyset$.

Example 3. Let $f(t) = t^{-1/2}$, $g(t) = 1 - t^2$. Then both the domain A_f and the range B_f of f are the interval $(0, \infty)$, the domain A_g of g is R and the range B_g is the interval $(-\infty, 1]$. Also, $A_f \cap B_g = (0, 1] = C$. It is easy to verify that the domain of the composite $f \circ g$ is the set $g^{-1}(C) = (-1, 1) \neq A_g$.

If f is a function defined on a set A and if $f(A) \subset A$, then f can be composed with itself: $(f \circ f)(t) = f(f(t))$, $(f \circ f \circ f)(t) = f(f(f(t)))$ and so on. Thus, we obtain the composites $f \circ f, f \circ f \circ f, \ldots$. These are termed the iterates of the function f. Their properties are studied throughout Chapter 3 and also in parts of Chapters 4 and 5.

Given two functions f and g, new functions may be constructed by arithmetical operations. The functions obtained are $f + g, f - g, f \cdot g, f/g$ and f^g. Their meaning is obvious. For example, the function $f + g$ is defined by $(f + g)(t) = f(t) + g(t)$, which makes the value of $f + g$ at t equal to the sum of the values of f and g at t. All these functions may be regarded as composites. In fact, the function $f + g$, for example, is obtained by composition of functions f and g with the function φ of two variables given by $\varphi(u, v) = u + v$.

Let us note that $f \circ f$ should not be confused with $f \cdot f$, or $f \circ g$ with $f \cdot g$. Therefore, we shall denote the iterates of a function f as follows: $f \circ f = f^2, f \circ f \circ f = f^3$ and so on, while the square of f, that is $f \cdot f$, will be denoted by $(f(t))^2$. Thus $\cos^2 \pi = \cos(\cos \pi) = \cos(-1)$, but $(\cos \pi)^2 = (-1)^2 = 1 \neq \cos(-1)$. This notation is practical rather than usual as we shall deal with iterations much more often than with powers of functions.

Now we recall another important concept. Let f be a function defined on a set A. If $x \neq y$ implies $f(x) \neq f(y)$ for any $x, y \in A$, then f is called an injective function. Put $f(A) = B$. Then for every $y \in B$ there exists exactly one $z \in A$ such that $f(z) = y$. This element z depends on y and we may write $z = \varphi(y)$. Then φ is a function defined on B and its range is A, that is, $\varphi: B \to A$. Moreover, $f \circ \varphi$ is the identity function on B, that is $(f \circ \varphi)(t) = t$ for all $t \in B$, and analogously $\varphi \circ f$ is the identity function on A.

The function φ is inverse to f. As there exists at most one such φ for a given function f, it is called the inverse function of f and denoted by f^{-1}. The following theorem summarises the above discussion.

Theorem 1

Let $f : A \to B$ be an injective mapping of A onto B, that is, let $f(A) = B$. Then there exists exactly one function $f^{-1} : B \to A$ which is an injective mapping of B onto A and satisfies

and
$$(f \circ f^{-1})(t) = t \qquad for\ all \quad t \in B$$
$$(f^{-1} \circ f)(t) = t \qquad for\ all \quad t \in A.$$

Observe that the graph of the inverse function of f is the reflection of the graph Γ_f of f in the line $y = x$. Some examples are $(e^t)^{-1} = \ln t$, $(\ln t)^{-1} = e^t$, $(x^2)^{-1} = x^{1/2}$ and so on. The domain and the range should be considered carefully.

We conclude this section by pointing out another important type of function, namely that defined on the set N_0 of all non-negative integers $0, 1, 2, \ldots$ with values in R. If $\varphi : N_0 \to R$ is such a function, we usually write φ_n, a_n or x_n instead of $\varphi(n)$ and we say that φ is a sequence (in this case, a sequence of real numbers). Then we write $\{\varphi_n\}_{n=0}^{\infty}$ instead of φ. An example of a sequence is $\{n^2 + 2n - 1\}_{n=0}^{\infty}$; here we have $a_n = n^2 + 2n - 1$ (a_n is referred to as the nth term of the sequence).

Readers will have observed that we use R to denote the set of all real numbers and that N_0 denotes the set of all non-negative integers. The set of all rational numbers will be denoted by Q. It is necessary to explain one more piece of notation. The symbol (a, b) represents the open interval with end points a and b, that is the set $\{x \in R; a < x < b\}$. We write $[a, b)$ for $\{x \in R; a \leqslant x < b\}$ and the intervals $(a, b]$ and $[a, b]$ are defined analogously.

Exercises

1.1 Sketch the graph of the function (1.1).

1.2 Find the domain of the function
$$f(t) = \log[1 - (t^2 - 3t + 2)^{1/2}].$$

1.3 Find all the maximal intervals on which the function defined in Exercise 1.2 is injective. What does the inverse function on each of those intervals look like?

1.2 CONTINUOUS FUNCTIONS

We come now to another important notion — that of continuity. We define the neighbourhood of a point (number) $x \in R$ to mean any open interval U centred at x. (More precisely, a symmetrical neighbourhood of x is an interval $(x - \delta, x + \delta)$ with $\delta > 0$. It is sometimes referred to as the δ-neighbourhood of x.)

Definition 2. We say that a function $f: R \to R$ is continuous at $x_0 \in R$ if, given any neighbourhood I of the point $y_0 = f(x_0)$, there exists a neighbourhood J of x_0 such that $f(J) \subset I$.

The last condition may be expressed by
$$f^{-1}(I) \supset J$$

that is, the pre-image of any neighbourhood of $y_0 = f(x_0)$ includes some neighbourhood of x_0.

Example 4. The functions $f(x) = 3x$ and $g(x) = x^2 - 1$ are continuous at every point. On the other hand, the function f given by equation (1.2) is not continuous at any rational number. Similarly, the function $h(x) = [x]$ (the integer part of x, giving $h(2) = h(2.35) = h(e) = 2$, $h(-1.7) = -2$ and so on) is discontinuous at every integer and is continuous at any other (non-integer) number. Draw the respective sketches.

To prove that a given function is continuous at a given point, or that it is discontinuous, is a more difficult task. Yet in some simple cases it is not very hard. For instance, the function $f(t) = 3t$ from Example 4 is indeed continuous at every point. To prove that, choose $x_0 \in R$. Any neighbourhood of the point $f(x_0) = 3x_0$ may be written as $I = (3x_0 - \varepsilon, 3x_0 + \varepsilon)$, where ε is any positive number. Then, however, $f^{-1}(I) = (x_0 - \varepsilon/3, x_0 + \varepsilon/3)$ is a neighbourhood of x_0.

With more complicated functions, however, the proof of continuity may be much more difficult. In general, it is easier to prove that a function is not continuous at a point than to prove that it is continuous at that point. (In the former case it is sufficient to find just one suitable neighbourhood, while in the latter all neighbourhoods must be checked.)

Definition 3. A function $f: R \to R$ is said to be continuous on a set $A \subset R$ if it is continuous at every point of A.

The above definition of continuity at a point and on a set can be extended to functions which are not defined on the whole set R. We say that a function f, having a domain D, is continuous at a point $x_0 \in D$ if the pre-image of any neighbourhood of $y_0 = f(x_0)$ includes some relative neighbourhood of x_0 with respect to D, that is $J \cap D$ for some neighbourhood J of x_0. (Draw a sketch.)

It is intuitively evident that an uninterrupted line in the graph of a function means continuity, a disrupted line means discontinuity. However, this is only true provided that the domain of the function is an interval of or the whole set R. If, for instance, A is the set of all non-zero numbers and $g: A \to R$ is defined by $g(x) = 1$ for all $x \in A$, then g is a continuous function, although its graph is interrupted at $x = 0$.

The following theorems may be useful in investigating continuity of functions. We state them without proofs.

Theorem 2

Let f be a function continuous on a set A and g a function continuous on a set B. Then each of the functions $f + g, f - g,$ $f.g, f/g$ and f^g is continuous on $A \cap B \cap D$, where D denotes the domain of the new (composite) function.

Theorem 3

Let $A \subset R$ be an interval and let $f: A \to B$ be a continuous injective mapping of A onto B. Then the inverse function $f^{-1}: B \to A$ is continuous on B.

Theorem 4

If $f: A \to B$ and $g: C \to D$ are continuous functions (that is, f is continuous on A and g is continuous on C) and if $B \subset C$, then the composite function $g \circ f: A \to D$ is continuous on A.

It is easy to verify that the functions $f(x) = x$ and $f(x) = c$ (c being a constant in R) are continuous and this implies, by Theorem 2, that all rational functions

$$f(x) = \frac{a_0 + a_1 x + \ldots + a_m x^m}{b_0 + b_1 x + \ldots + b_n x^n}$$

are continuous on their domains (that is, at all the points where the denominator does not vanish). Also, the function $f(x) = x^n$, for every positive integer n, is continuous and if we consider it as mapping $[0, \infty)$ onto $[0, \infty]$, it is injective as well. Therefore, its inverse function $g(x) = x^{1/n}$ is a continuous mapping of $[0, \infty)$ onto $[0, \infty)$. If n is odd, this function may be defined on the whole set R. Using the above results and Theorem 4, we can easily verify that every function $f(x) = (a_0 + a_1 x + \ldots + a_m x^m)^{1/n}$ is continuous on its domain.

It is a little more difficult to prove that other elementary functions, namely $\sin x$, $\cos x$, $\cot x$, $\log_a x$ and a^x, are continuous on their maximal domains. However, roughly speaking, every function constructed from elementary functions in an admissible way (that is, as indicated in Theorems 2, 3 and 4) is continuous on its domain.

Example 5. Functions

$$\frac{\sin (x^2 + 1)^{1/2}}{\log 3x}, \qquad \tan (x^2/2), \qquad e^{1/x}$$

are continuous on the sets $(0, 1/3) \cup (1/3, \infty)$, $R \setminus \{\pm\sqrt{\pi}, \pm\sqrt{3\pi}, \pm\sqrt{5\pi}, \ldots\}$ and $R \setminus \{0\}$, respectively.

Note that all the above statements are true under the assumption that by the maximal domain we mean the 'usual' domain, which is the set of all those x that may be substituted in the respective propositional form and will not give a term of the type $1/0$ or a similar nonsensical expression. For example, the usual domain of the function $f(x) = 1/x$ is $(-\infty, 0) \cup (0, \infty)$. Of course, we could extend this function to obtain a function F defined on the whole of R by putting $F(0) = 0$ for instance. But this function F is not continuous at zero. In fact, the function f cannot be extended in any way to a function with domain R and continuous at zero. (Why?)

Theorem 5

Let f be a function continuous on A and let $I \subset A$ be an interval. Then f(I) is an interval. (In other words, a continuous image of an interval is again an interval.) Moreover, if I is a bounded, closed interval, then f(I) is also a bounded and closed interval.

Exercises

1.4 If f is a continuous function (or a function continuous at a point x), then the function $|f|$ is also continuous (at x). Prove that this is true. Is the converse statement true?

1.5 (a) Prove that the functions $\tan x$, $\cot x$ and $1/x$ cannot be extended to become defined and continuous on the whole of R.
(b) Prove that this is also true for the rational functions of the type

$$\frac{1}{(x - a_1)(x - a_2) \dots (x - a_n)}.$$

1.6 Show that the function defined by (1.2) is continuous at every irrational number.

1.7 Prove Theorem 4.

1.8 Using Theorem 5, prove the following important proposition: if f is a function continuous on $[a, b]$ and if $f(a) < 0$, $f(b) > 0$, then $f(c) = 0$ for some $c \in (a, b)$.

1.3 DERIVATIVE OF A FUNCTION

Let $f(t)$ be a continuous function defined on an interval $I = (a, b)$ and let $c \in (a, b)$. Consider the function

$$F(t) = \frac{f(t) - f(c)}{t - c}.$$

This function is defined in a 'natural' way on the whole of *I*, except at the point *c*. It has an evident interpretation, namely the slope of the straight line connecting the points $[t, f(t)]$ and $[c, f(c)]$ of the graph of *f*, that is the tangent of the angle formed by the said straight line and the *x* axis. Moreover, from Theorem 2, the function *F* is continuous on its domain, which is the set $(a, c) \cup (c, b)$ (see figure 1.1).

   If the function *F* can be assigned a value at *c* in such a way that the function thus extended is continuous on the whole interval *I*, then that value $F(c)$ is termed the derivative of *f* at the point *c* and is denoted by $f'(c)$. The geometric interpretation of the derivative is evident. It is the slope of the tangent to the graph of *f* at *c* (see figure 1.1).

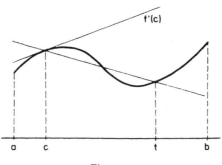

Figure 1.1

   The problem is whether the derivative exists. First of all we show that it need not exist in general.
   Example 6. Let $f(t) = |t|$ for all $t \in R$. Put $c = 0$. Then, for $t > 0$,

$$F(t) = \frac{|t| - 0}{t - 0} = \frac{t - 0}{t - 0} = 1,$$

while for $t < 0$

$$F(t) = \frac{-t - 0}{t - 0} = -1.$$

Thus F attains two values, 1 and -1, in every neighbourhood of zero. Therefore it cannot be extended to the point $c = 0$ to obtain a continuous function. The situation is illustrated by figures 1.2a and 1.2b showing, respectively, the functions f and F. Therefore, the function $f(t) = |t|$ has no derivative at zero.

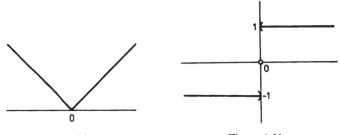

Figure 1.2a Figure 1.2b

Example 7. Consider the function f defined by

$$f(t) = \begin{cases} t \sin(1/t) & \text{for} \quad t \neq 0 \\ 0 & \text{for} \quad t = 0. \end{cases}$$

This function is continuous everywhere, even at zero, in spite of the fact that the point zero does not belong to the natural domain of $\sin(1/t)$; the graph of f is sketched in figure 1.3a. Again, put $c = 0$ and examine the value at c of the function

$$F(t) = \frac{f(t) - f(0)}{t - 0}.$$

To solve, at

$$t_k = \left(\frac{\pi}{2} + 2k\pi \right)^{-1}, \quad k = 1, 2, 3, \ldots$$

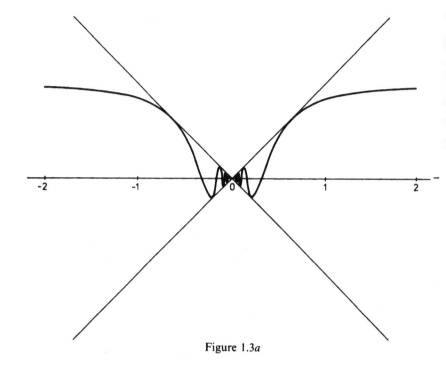

Figure 1.3a

we have

$$f(t_k) = t_k \sin(1/t_k) = t_k$$

while at

$$s_k = \left(-\frac{\pi}{2} + 2k\pi\right)^{-1}, \quad k = 1, 2, 3, \ldots$$

we get

$$f(s_k) = s_k \sin(1/s_k) = -s_k.$$

Thus $F(t_k) = 1$ and $F(s_k) = -1$ for all k. If U is any neighbourhood of zero, it always contains some t_k and some s_k values. Therefore F attains, in every neighbourhood of zero, the values 1 and -1 (a stronger proposition is true, i.e. F

attains in U all the values between -1 and 1), hence it cannot be continuous. The graph of F is shown in figure 1.3b.

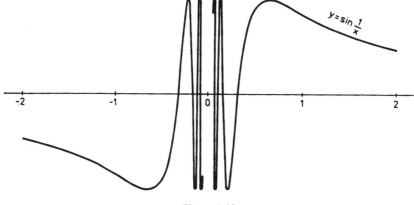

Figure 1.3b

However, some functions do have a derivative.

Example 8. Let $f(t) = t^2$ and let $c \in R$ be an arbitrary point. Then

$$F(t) = \frac{f(t) - f(c)}{t - c} = \frac{t^2 - c^2}{t - c} = t + c$$

for $t \neq c$. The graph of $F(t)$ is depicted in figure 1.4. Evidently, by putting $F(c) = c + c$ we obtain the continuity of F at c. Thus the function $f(t) = t^2$ has, at every point t, the derivative $f'(t) = 2t$.

If we continued further with similar considerations, we would discover the rules which are summed up in the following theorem.

Theorem 6

Each of the following functions has a derivative at every point of its domain, namely

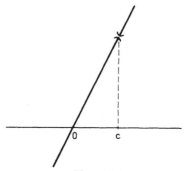

Figure 1.4

1. $c' = 0$ (*c is a constant*),
2. $(t^a)' = at^{a-1}$ (*a is a constant, $a \neq 0$*),
3. $(\sin t)' = \cos t$,
4. $(\cos t)' = -\sin t$,
5. $(e^t)' = e^t$,
6. $(\ln t)' = 1/t$.

Observe that the function e^t is not changed by derivation.

In order to be able to compute the derivatives of more complicated functions, we introduce additional rules.

Theorem 7

Derivation of composite functions obeys the following rules:

1. $(f \pm g)' = f' \pm g'$
2. $(f \cdot g)' = f' \cdot g + f \cdot g'$
3. $\left(\dfrac{f}{g}\right)' = \dfrac{f' \cdot g - f \cdot g'}{g \cdot g}$
4. $(f \circ g)' = (f' \circ g) g'$
   (*more precisely, $(f(g(t)))' = f'(g(t)) \cdot g'(t)$*).

In all the above cases, the derivative exists on every interval on which both f and g have derivatives and the respective composite function is defined.

With the rules listed above, we can differentiate all the functions composed from the elementary functions in an admissible way.

Example 9.

1. $(3t^4)' = 12t^3$ (we have used rule 2 of Theorem 6 and rule 2 of Theorem 7);

2. $(x^2 \sin x)' = 2x \sin x + x^2 \cos x$ (using rules 2 and 3 of Theorem 6 and rule 2 of Theorem 7);

3. $(\tan t)' = \left(\dfrac{\sin t}{\cos t} \right)' = \dfrac{(\sin t)^2 + (\cos t)^2}{(\cos t)^2} = (\cos t)^{-2}$

(using rules 3 and 4 of Theorem 6 and rule 3 of Theorem 7);

4. $(a')' = (e^{t \ln a})' = (e^{t \ln a})(t \ln a)' = (e^{t \ln a}) \ln a =$
$= a' \ln a$ (rules 5, 2 and 1 of Theorem 6 and rules 2 and 4 of Theorem 7 have been used);

5. $[\sin (t^2 + 1)^{1/2}]' = [\cos (t^2 + 1)^{1/2}][(t^2 + 1)^{1/2}]'$

$$= [\cos (t^2 + 1)^{1/2}] \tfrac{1}{2} (t^2 + 1)^{-1/2}(t^2 + 1)'$$

$$= [\cos (t^2 + 1)^{1/2}] \tfrac{1}{2} (t^2 + 1)^{-1/2}(2t)$$

(using rule 4 of Theorem 7 twice, rules 2 and 3 of Theorem 6, rule 1 of Theorem 7 and again rule 2 of Theorem 6).

The above rules can also be used for solving more practical problems.

Example 10. Write an equation for the tangent to the curve (that is, the graph of the function) $y = \sin x$ at the point $[x_0, y_0] = [\pi, 0]$.

Solution: We know that the derivative of a function f at x_0 is equal to the slope of the tangent at $[x_0, f(x_0)]$. In our case, we obtain the slope $k = (\sin x)'_{x = \pi} = \cos \pi = -1$. Hence the tangent at $[\pi, 0]$ is described by the equation

$$y = k(x - x_0) + y_0 = -x + \pi.$$

The following theorem can be proved.

:

Theorem 8

Suppose that f has a derivative f' on a whole interval I. If, for every t ∈ I, we have

    1. $f'(t) < 0$,   *then f is decreasing on I,*
    2. $f'(t) > 0$,   *then f is increasing on I.*

    The above theorem will be frequently used for investigating monotonicity of functions.

    Example 11. Let $f(t) = 2t^3 + 3t^2 - 12t + 8$. We would like to know the number of zeros (i.e. points x such that $f(x) = 0$) of the function f.

Solution: Calculate the derivative

$$f'(t) = 6t^2 + 6t - 12 = 6(t - 1)(t + 2).$$

It is easy to check that

$$f'(t) < 0 \qquad \text{for} \quad t \in (-2, 1)$$

and

$$f'(t) > 0 \qquad \text{for} \quad t \in (-\infty, -2) \cup (1, \infty).$$

    Thus the function f is increasing on the interval $(-\infty, -2)$ as well as on $(1, \infty)$. (Of course, this does not mean that it is increasing on $(-\infty, -2) \cup (1, \infty)$!) Further, f is decreasing on $(-2, 1)$. On the other hand, $f(-2) = 36$, $f(1) = 1$, and since f is decreasing on $(-2, 1)$ and increasing on $(1, \infty)$, it is evidently positive over the whole interval $(-2, \infty)$ and cannot have a zero there. We observe also that $f(-4) = -24$ and hence f has a zero in the interval $(-4, -2)$ (see Exercise 1.8). It is the only zero of f because f is increasing on $(-\infty, -2)$ and its graph cannot intersect the x axis more than once. (Sketch the graph of f.)

Exercises

1.9 Compute the derivatives of the following functions:

1. $(t^2 + 3t + 5)/(\sin t)^2$
2. $[t + (t + t^{1/2})^{1/2}]^{1/2}$.

1.10 Find the number of zeros of the following functions and sketch their graphs:

1. $t^3 + t^2 + t + 1$
2. $2t^3 - 3t^2 + 4t - 5$
3. $(t + 1)(t^2 + 1)^{-1/2}$

2. FUNCTIONAL EQUATIONS IN SEVERAL VARIABLES

2.1 INTRODUCTION

Let us begin with an example. A group of n anglers are sitting around a pool. Each of them is fishing with his own fishing rod. Assume that everybody will be happy to catch just one fish. However, none of them will go away without a catch. It is evident that if there are enough fish in the pool, the number of fishermen sitting will decrease with time. We are interested to know how the decreasing number of anglers can be expressed as a function of time.

This task looks quite difficult at first sight. We will show presently that it is not really that difficult. First of all, we introduce some notation to make the problem clearer. We begin to measure the time from the moment when the anglers started fishing. Let $f(t)$ denote the relationship between the number of anglers still fishing at time t and the total number of fishermen n. Then the number of anglers having caught nothing by the time t will be $nf(t)$. We have to determine the function f.

What do we know about f so far? Fish are abundant and we assume that the anglers do not disturb each other while fishing. Also, the waiting time has no influence on the probability of catching a fish in the next moment, i.e. a fisherman who has been waiting for a catch in vain for several hours has the same chance of catching something in the next moment as the angler who has just arrived. In short, we can say that:

1. the function f is independent of n,
2. the function f does not depend on the choice of the initial moment of time.

Thus, the number of fishermen still sitting near the pool at the time $t + h$ can be expressed in two ways: either by $nf(t + h)$ or by $(nf(t))f(h)$. Figure 2.1 illustrates this. We have now obtained the equation $nf(t + h) = nf(t)f(h)$, or

$$f(t + h) = f(t)f(h). \qquad (2.1)$$

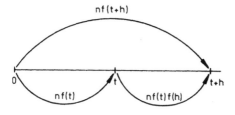

Figure 2.1

In this equation the function f is still unknown. Such an equation is termed a functional equation.

The assumptions could have been formulated in a different way. Instead of 1 and 2 above, it would have been sufficient to suppose that:

3. in equally long time intervals, the ratio between the number of anglers at the beginning of the interval and that at its end is constant.

Thus if $[0, h]$ and $[t, t + h]$ are two intervals having the same length h, we have

$$\frac{n}{nf(h)} = \frac{nf(t)}{nf(t + h)}. \qquad (2.2)$$

In fact, on the left-hand side of the above equation we have the ratio between the numbers of anglers at the time zero and at the time h, while the right-hand side expresses the ratio of the numbers of fishermen at the times t and $t + h$. Equation (2.2) can be easily reduced to yield equation (2.1).

The problem concerning the fishermen will be solved as soon as we find a function f satisfying equation (2.1). Therefore, in what follows below we shall study the methods for solving such equations, and in the subsequent sections we shall gradually give some more complicated equations and also examples of their applications.

An important concept we shall deal with is that of a function. For a reader who has not read the introductory chapter, we recall that by a function we always mean a real (i. e. real-valued) function f defined on a subset D of the set R of real numbers (that is, for every $x \in D$ there is exactly one number $f(x)$). The set D will be referred to as the domain of f.

We are not going to define exactly what a functional equation is, as such a definition is not simple. Instead, we give below some more examples of functional equations, which may help to give a more precise idea of the meaning of that concept:

$$f(xy) = xf(y) \tag{2.3}$$

$$f(xy) = f(x) + f(y) \tag{2.4}$$

$$f(x + y) + f(x - y) = 2f(x)f(y) \tag{2.5}$$

$$f(xy) = f(x/y) \tag{2.6}$$

$$f(3t) = t^2 f(t) \tag{2.7}$$

$$f(t^2 - 2f(t)) = f(f(t)) \tag{2.8}$$

$$f(f(f(t))) = t \qquad (2.9)$$

$$f'(t) = t^3 \qquad (2.10)$$

$$f'(t) = f(t) + 4t^2 \qquad (2.11)$$

(the term f' denotes the derivative of f). In each of the above equations, f is the unknown function. Equations (2.10) and (2.11) differ from the preceding ones by the occurrence of the derivative f' besides the unknown function f itself. Such equations are termed differential equations. However, we shall not discuss them in this book although readers may easily check that any function defined on the set R of reals which satisfies equation (2.10) has the form $f(x) = (x^4/4) + c$, where c is an arbitrary constant.

The remaining equations can be classified into two groups. Equations (2.3) to (2.6) are equations in two variables x and y — such equations will be the object of our study in this chapter. Equations (2.7), (2.8) and (2.9) are equations in one variable t; to solve them is, in general, more difficult than to solve equations in two (or more) variables and we shall study them in Chapter 5.

To solve a functional equation means finding all the functions satisfying the given relation. Sometimes, the domain of the functions is also prescribed. The solution of the equation is, of course, influenced by such a condition.

Example 1. Find all the functions f defined on the set R of reals that satisfy the functional equation (2.4).

How shall we solve this problem? It can be done in a quite natural way. Assume first that the equation has a solution f and from that assumption we shall deduce (if possible) what form the solution must have. In other words, we shall find the conditions necessary for f to be a solution of the equation. Then, however, we must verify that the function obtained does solve the equation (see Example 4).

Thus, suppose that f is a solution of equation (2.4). Since the domain of f is the whole set R, (2.4) holds for any two real numbers x and y; in particular, for $x = 0$ and an arbitrary y. Hence

$$f(0) = f(0) + f(y).$$

Subtracting $f(0)$ from both sides of the above equation, we get $f(y) = 0$. This must be true for every real y. Therefore, whenever a function defined over the whole set R solves equation (2.4), it equals zero at every $x \in R$. It is easy to see that this function is really a solution of equation (2.4). We have just proved that the function defined by $f(x) = 0$ at every $x \in R$ is the unique function defined on R which solves (2.4).

Let us stay for a while with equation (2.4), and look for solutions whose domain is the set $R \setminus \{0\}$ of all real numbers different from zero. It is easy to verify that in this case the equation has at least two solutions. Besides the zero function, it also has another solution, the function h, defined on $R \setminus \{0\}$ by $h(x) = \log |x|$.

We have seen that determining the domain of the solution of a functional equation is very important. Therefore, it will be convenient to agree on the following terminology. If we are looking for solutions of a functional equation that are defined on a set D, we shall say that we seek the solutions of that equation in the set D or in the domain D. Recall that two functions are equal if they have the same domain and the same values at every point of their common domain.

In Example 1 we have seen that the set of all solutions of equation (2.4) in the 'smaller' domain $R \setminus \{0\}$ has more elements than the set of all its solutions in the 'larger' domain R. It might seem that, for any functional equation, the following is true: if D_1 and D_2 are sets of real numbers and $D_1 \subset D_2$, then the set of solutions of that equation in the domain D_2 is not 'richer' than the set of all solutions of the same equation in the domain D_1. But this is not true in general (consider the solutions of (2.4) in the domain $D_3 = \{1\}$).

Solving some functional equations is easy, but there are equations that are very difficult to solve. We cannot expect to solve easily any functional equation we write down. Equation (2.5) is an example of this type of equation; one of its solutions is the function $\cos x$. In the following two examples, equations (2.3) and (2.6) are solved.

Example 2. Find all the solutions of the functional equation (2.3) in the domain R of real numbers.

Solution: Suppose that f_0 is a solution of (2.3), defined on R. Putting $y = 1$, we have $f_0(x) = xf_0(1)$ for every $x \in R$. Thus a necessary condition for f_0 to solve (2.3) is that f_0 be the function ax where a is a real constant. We must still verify that every function ax does solve the equation (2.3). This, however, is not difficult: $f_0(xy) = a(xy)$, $xf_0(y) = x(ay)$, hence $f_0(xy) = xf_0(y)$.

Example 3. Find all the solutions of the functional equation (2.6) in the domain R^+ of all positive reals.

Solution: Let f be a solution of (2.6). The equation is true also for $x = y$. Substituting, we get $f(x^2) = f(x/x) = f(1)$, therefore f assumes the value $f(1)$ at any such number y in the domain R^+ of f which is the square of some positive number. Of course, every positive real y has the said property. Thus, whenever a function solves (2.6) in R^+, it is a constant function. The converse assertion is evidently true, too.

The above three examples may have given us some idea of how to solve functional equations. We always try to choose special numbers for x and y, such as $0, 1, -1, x = y$ or $x = -y$. Of course, it is difficult to give a general rule. The method always depends on the given functional equation.

Example 4. Find all the functions defined on the set R of real numbers that solve the functional equation

$$2f(u + v) = f(u + 2v) + u^3. \tag{2.12}$$

Solution: Assume that f_0 is a solution of (2.12). Putting $v = 0$, we get $2f_0(u) = f_0(u) + u^3$, that is, $f_0(u) = u^3$. Therefore, if a function defined on R solves (2.12), then it is the function $f_0(u) = u^3$ for all $u \in R$.

Let us now check whether f_0 really solves the equation (2.12). The left-hand side reduces to $2f_0(u + v) = 2(u + v)^3 = L(u, v)$; the right-hand side gives $f_0(u + 2v) + u^3 = (u + 2v)^3 + u^3 = P(u, v)$. The function f_0 will be a solution of our equation if, and only if, $L(u, v) = P(u, v)$ for every $u, v \in R$. However, it is easy to see that this equality is not always true. For example, by putting $u = 0$ we obtain $L(0, v) = 2v^3 \neq 8v^3 = P(0, v)$ for any $v \neq 0$. Therefore, equation (2.12) has no solution in R.

There are many examples similar to Examples 2, 3 and 4. Some of them are given below as exercises. We suggest that readers do at least some of them before continuing.

Exercises

2.1 In the domain R of all real numbers, solve the following functional equations:

1. $f(xy) = f(x)$;
2. $f(x + y) = f(x)$;
3. $f(x + y) = f(x)e^y$.

2.2 Find all the solutions of the following functional equations defined on the set R^+ of all positive reals:

1. $f(xy) = f(x)y^a$ for all $x, y \in R^+$, a is a real constant;
2. $f(x^y) = f(x)y$ for all $x \in R^+$, $y \in R$;
*3. $f(x^y) = f(x)$ for all $x, y \in R^+$.

2.3 Find all the solutions in the domain R of the functional equations:

1. $f(x + y) + f(x - y) = f(x) + 6xy[f(y)]^{1/3} + x^3$;
2. $f(x + y) + f(x + 2y) = f(x) + 2xy + x^3$;
*3. $f(x + y) - 2f(x - y) + f(x) + f(y) = 4y + 1$.

2.4 Check that $\cos x$ is a solution of equation (2.5). Write a similar equation for $\tan x$.

2.2 THE CAUCHY FUNCTIONAL EQUATION AND RELATED EQUATIONS

We shall now deal with the functional equation

$$f(x + y) = f(x) + f(y) \qquad (2.13)$$

which has a privileged position in mathematics. It is encountered in almost all mathematical disciplines. Its properties are studied not only in numerical domains, as we shall do here, but also in various abstract sets, such as groups, systems of functions etc*. This equation is not as easy to solve as the preceding ones. The method we shall give here was first used in 1821 by the French mathematician A L Cauchy, and the equation is named after him.

Suppose, as usual, that the equation has a solution f in the domain R. Putting $x = y = 0$, we have $f(0) = = f(0) + f(0)$, hence $f(0) = 0$. If $x = y$, we get $f(x + x) = = f(x) + f(x)$, therefore $f(2x) = 2f(x)$ for every $x \in R$. Analogously, $f(3x) = f(2x + x) = f(2x) + f(x) = 2f(x) + + f(x) = 3f(x), f(4x) = 4f(x)$, and so on. This leads us to propose that for each positive integer t and for every $x \in R$,

$$f(tx) = tf(x). \qquad (2.14)$$

---

* Monographs [8] and [17] cited in the bibliography at the end of this book are devoted mainly to this equation. See also [1] and [2].

We shall try to prove this using mathematical induction. Clearly, (2.14) holds for $t = 1$. Let us assume that (2.14) is true for $t = k$ and we shall show its validity also for $t = k + 1$. In fact, $f((k + 1)x) = f(kx + x) = f(kx) + f(x) = kf(x) + f(x) = (k + 1)f(x)$ (in the second equality we made use of (2.13) and in the third we used the induction hypothesis). Thus (2.14) is true for every positive integer t.

Now, let m and n be positive integers. Substitute in (2.14) $t = n$, $x = m/n$. We get

$$f(m) = f\left(n\,\frac{m}{n}\right) = nf\left(\frac{m}{n}\right)$$

which easily yields

$$f\left(\frac{m}{n}\right) = \frac{1}{n} f(m) = \frac{1}{n} f(m.1) = \frac{m}{n} f(1) \qquad (2.15)$$

(in the third equality we again made use of (2.14)). Since for every $x \in R$ we have

$$0 = f(x - x) = f(x) + f(-x)$$

it follows that

$$f(x) = -f(-x) \qquad x \in R.$$

This, together with (2.15), implies that we have established the following theorem which will prove very useful later.

Theorem 1

Let a function f defined on the set R of all real numbers be a solution of the Cauchy functional equation (2.13). Then every rational number r satisfies

$$f(r) = rf(1). \qquad (2.16)$$

The above theorem implies, besides other things, that if we know the value of f at 1, we know its value at any rational point. Unfortunately, from the knowledge that f solves the Cauchy equation we cannot deduce any reasonable condition relating the values of f at irrational points with its value at 1. It is evident, and can be checked by substitution, that the function f defined by (2.16) for every real r is a solution of (2.13), but it is not true that every solution of our equation is given by (2.16) — we shall show this in § 2.5.

However, if a function f which solves the Cauchy equation is assumed to have some appropriate additional property, then (2.16) will be true for all $r \in R$. Monotonicity is an example of such a property. Let us recall this concept. We say that a function f is non-decreasing if for any two points $x < y$ in its domain we have $f(x) \leqslant f(y)$. Analogously, f is said to be non-increasing if $x < y$ implies $f(x) \geqslant f(y)$. Finally, f is termed monotone if it is either non-increasing or non-decreasing.

Let us try to find out what all the monotone solutions of (2.13) look like. Let f be such a solution. We now proceed to show that (2.16) holds for every real r.

Assume the contrary, namely that for some $x \in R$ the relationship $f(x) \neq xf(1)$ is true. Then, either $f(x) > xf(1)$ or $f(x) < xf(1)$. In the former case we can find rational

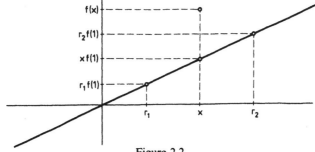

Figure 2.2

numbers r_1 and r_2 such that $r_1 < x < r_2$ and both $r_1 f(1) < f(x)$ and $r_2 f(1) < f(x)$ (see figure 2.2). Then from equation (2.16) we have $f(r_1) < f(x)$ and $f(r_2) < f(x)$; hence f is not monotone. Thus, the assumption that (2.16) is false for some $r \in R$ yields a contradiction. In the case $f(x) < xf(1)$ we proceed analogously.

Thus every monotone solution of (2.13) necessarily has the form $f(x) = ax$, where a is a constant (equal to $f(1)$). Conversely, every function ax solves the considered equation. We have thus proved the following theorem.

Theorem 2

A monotone function f defined on R solves the Cauchy equation (2.13) if and only if $f(x) = ax$ for every $x \in R$, where a is a real constant.

Methods like those used in proving the above two theorems may be applied also in the following example.

Example 5. Find all the monotone solutions of equation (2.1) defined on the set R^+ of positive real numbers.
Solution: Let f be a monotone solution of (2.1). Since for every $x \in R^+$ we have

$$f(x) = f(x/2 + x/2) = (f(x/2))^2 \geqslant 0,$$

the function f is non-negative. From (2.1) we get $f(2x) = f(x)f(x) = (f(x))^2$, $f(3x) = (f(x))^3$ etc. It can be proved by induction that

$$f(nx) = (f(x))^n \qquad (2.17)$$

for every positive integer n (we recommend readers write down the induction proof in detail). Putting $x = 1/n$ in (2.17), we get

$$f(1) = f(n(1/n)) = (f(1/n))^n$$

and hence

$$f(1/n) = (f(1))^{1/n}. \qquad (2.18)$$

If m and n are positive integers, (2.17) and (2.18) imply

$$f(m/n) = f(m(1/n)) = (f(1/n))^m = ((f(1))^{1/n})^m = (f(1))^{m/n}.$$

Consequently, for any rational number r we have

$$f(r) = (f(1))^r. \qquad (2.19)$$

The monotonicity condition for f implies that (2.19) holds true for every positive real r (sketch a diagram of this).

Thus, if f is a monotone solution of (2.1), then f is a function defined at every $r \in R^+$ by (2.19). Conversely, every such function is a solution of (2.1). Therefore, all the monotone functions solving (2.1) on R^+ are exactly all the functions of the form $f(x) = a^x$, where a is a non-negative, real number. Many other functional equations can be solved analogously. In some cases, Theorem 2 may be applied directly, as we shall see in the following example.

Example 6. Let us find all the monotone functions f defined on the set R^+ of positive reals and satisfying (2.4), i.e.

$$f(xy) = f(x) + f(y).$$

Solution: Let f be a solution of equation (2.4). Define a new function F on R by $F(t) = f(e^t)$. Since both e^t and f are monotone functions, F is monotone as well. In fact, let f be, say, a non-decreasing function and let $t < s$. Then $e^t < e^s$ and hence $F(t) = f(e^t) \leqslant f(e^s) = F(s)$. The remaining cases are left to the readers. For any two numbers $s, t \in R$ we have

$$F(s + t) = f(e^{(s + t)}) = f(e^s e^t) = f(e^s) + f(e^t)$$
$$= F(s) + F(t)$$

thus F is a monotone solution of the functional equation

(2.13). Hence, by Theorem 2, we have $F(t) = at$ for every $t \in R$. So $f(e^t) = at$ and, therefore, (denoting e^t by y) for all $y \in R^+$ we have (why?)

$$f(y) = a \ln y. \tag{2.20}$$

Conversely, every function of the form (2.20) is a solution of equation (2.4).

Let us look again at the way we proved Theorem 2. We had a function f with prescribed values on the set Q of rational numbers, and for each $r \in Q$ we had $f(r) = rf(1)$. Using this fact and the monotonicity of f we were able to determine the value of f at any real number. Note that we did not make use of the fact that f was a solution of an equation. In other words, given a function h defined on Q by $h(r) = rf(1)$, there exists exactly one monotone function f defined on R and satisfying $f(r) = h(r)$ for every $r \in Q$. In such a case we say that the function f defined on R is an extension of the function h defined on the 'smaller' set Q^*. The following example shows that not every function defined on Q shares this property with h.

Example 7. Let g be a function defined on the set Q of rationals by $g(x) = 0$ for $x \in Q$, $x < 2^{1/2}$, and $g(x) = 1$ for $x \in Q$, $x \geq 2^{1/2}$ (as we know, $2^{1/2}$ is not a rational number, so we may write the strict inequality $x > 2^{1/2}$ instead of the latter one). Evidently, the function g is monotone, namely non-decreasing. However, there are infinitely many monotone functions w defined on R and satisfying $g(x) = w(x)$ for each $x \in Q$; one of them is shown in figure 2.3. Such a function w must satisfy $w(x) = 0$ for $x < 2^{1/2}$ and $w(x) = 1$ for $x > 2^{1/2}$, but $w(2^{1/2})$ may be any number between 0 and 1.

---

* An exact definition of this reads as follows. Let D_1 be the domain of a function f_1, let D_2 be the domain of a function f_2. Also, let $D_1 \subset D_2$ and $f_1(x) = f_2(x)$ for every $x \in D_1$. Then we say that f_2 is an extension of f_1.

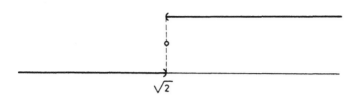

Figure 2.3

The reason why g has infinitely many monotone extensions is evident: no monotone extension w of g is continuous at $2^{1/2}$, which is the point where the uniqueness of w breaks down.

The above example leads us to conjecture that continuous extensions of 'reasonable' functions, if they exist, are uniquely determined. We are going to prove that this is really the case. First, however, we specify some notation.

Let $a \in R$. Then a neighbourhood of a is any open interval with its centre at a. If $A \subset B$ are arbitrary sets of real numbers, we say that A is dense in B if every neighbourhood of any point of B contains some point of A. Thus, for example, the set of all rational numbers is dense in R, but, also, the set of irrationals or the set of all numbers of the form $m\,2^{-n}$, where m is an integer and n a positive integer, is dense in R. On the other hand, the only dense subset of the set Z of integers is the whole set Z. Now we can prove the following important theorem.

Theorem 3

Let f, g be two continuous functions defined on a set A. Let B be a dense subset of A and suppose $f(x) = g(x)$ for each $x \in B$. Then $f(x) = g(x)$ for every $x \in A$, that is, $f = g$.

In other words, two continuous functions having the same domain are equal whenever they assume equal values on a dense subset of their domain.

This theorem can be proved as follows. We suggest readers begin by drawing a diagram. Now we make an exact proof by contradiction. Suppose that f and g are continuous functions on A having the same values on B, and that there exists a point x contained in A, but not in B, at which $f(x) \neq g(x)$. We shall assume that $f(x) > g(x)$ (in the case of the converse inequality, the proof is analogous). Let

$$f(x) - g(x) = a > 0. \qquad (2.21)$$

Since the function f is continuous at x, there is a neighbourhood $O_{f(x)}$ of x such that $|f(x) - f(y)| < a/2$ for every $y \in A \cap O_{f(x)}$. Similarly, for some neighbourhood $O_{g(x)}$ of x we have $|g(x) - g(y)| < a/2$ at every $y \in A \cap O_{g(x)}$. Since B is dense in A, there exists a point z in B contained in $A \cap O_{f(x)} \cap O_{g(x)}$. Because $f(z) = g(z)$, we have

$$|f(x) - g(x)| = |f(x) - f(z) + f(z) - g(x)|$$
$$= |f(x) - f(z) + g(z) - g(x)|$$
$$\leqslant |f(x) - f(z)| + |g(z) - g(x)|$$
$$< a/2 + a/2 = a$$

which contradicts (2.21). Theorem 3 is proved.

Note, the proof can be simplified if we use the fact that the difference of any two continuous functions is again a continuous function and examine the function $h(x) = = f(x) - g(x)$, which equals zero on B and attains at least one non-zero value on A. Obviously, such a function h cannot be continuous.

Theorem 3 has several interesting corollaries. For instance, it may prove useful when seeking continuous solutions of functional equations.

Example 8. Find all the continuous solutions of the functional equation $f(x + y) = f(x) + f(y)$ defined on the set R of real numbers.

Solution: Let f be a continuous solution of the given equation in R. From Theorem 1, for every rational number $r \in Q$ we have $f(r) = rf(1)$. The function g, defined by $g(x) = = xf(1)$ for all $x \in R$, is continuous and has the same value as f at every point of Q. Since Q is a dense subset of R, Theorem 3 implies $f = g$. Thus $f(x) = xf(1)$ for every $x \in R$.

Conversely, every function of the form ax is a continuous solution of the Cauchy functional equation. Therefore the continuous solutions of (2.13) in R are exactly the functions ax, where a is a real constant.

To conclude this section, we present two other interesting results concerning the Cauchy functional equation. It is found that the solutions of the said equation defined on R have the form ax under much weaker assumptions than continuity or monotonicity on the whole set R.

Theorem 4

Let f be a solution in R of the Cauchy functional equation which is upper bounded on some interval $I \subset R$. Then f is a continuous solution (of the form $f(x) = xf(1)$ for all $x \in R$).

The following proof is a little more difficult than the preceding ones and may be skipped. The upper boundedness of f on an interval I means that for some constant M we have $f(x) < M$ for any $x \in I$. Assume that the interval I is also bounded (if not, consider a suitable subinterval instead). Then the function

$$g(x) = f(x) - xf(1)$$

is upper bounded on I as well, in general, by another constant M' (why?). It is easy to verify that g solves the Cauchy equation and, moreover, from Theorem 1, for every rational number r we have

$$g(r) = f(r) - rf(1) = rf(1) - rf(1) = 0.$$

Given any real number y, we can find a rational number r such that $y + r = x$ is in I. Then, however,

$$g(x) = g(y + r) = g(y) + g(r) = g(y)$$

and, because $g(x) < M'$, we get $g(y) < M'$. We see that g is a solution of the Cauchy equation, which is upper bounded by M' on the whole set R. We complete the proof by showing that this implies $g(x) = 0$ (that is, $f(x) = xf(1)$) for all $x \in R$.

Suppose that $g(x_0) = a \neq 0$ for some $x_0 \in R$. We may assume $a > 0$ (otherwise we take $-x_0$ instead of x_0; why?). Then for every positive integer n we have

$$g(nx_0) = ng(x_0) = na$$

and, since g is upper bounded, we should have $na < M'$. This inequality is false, of course, for sufficiently large n. Thus g is not upper bounded, a contradiction.

Theorem 4 immediately implies the following result.

Theorem 5

If f is a solution in R of the Cauchy equation, continuous at some point x_0, then f is continuous everywhere on R (and hence it is the function $f(x) = xf(1)$).

Proof is obvious. If f is continuous at x_0 and x is near x_0, i.e. if x is in a suitable neighbourhood (interval) U of x_0, then $f(x)$ is close to $f(x_0)$, that is

$$f(x) < f(x_0) + \varepsilon$$

for some positive ε. Thus, f is upper bounded on U and Theorem 4 applies.

The methods presented in this section may also be used to solve other functional equations. Some of them are given in the following exercises.

Exercises

2.5 Find all the solutions of the Cauchy functional equation (2.13) in the sets:

1. $Q\,2^{1/2}$ of all numbers of the form $r\,2^{1/2}$, where r is a rational number;
*2. $Q + Q\,2^{1/2}$ of all numbers of the form $r + s\,2^{1/2}$, where r and s are rational.
Which of these solutions are continuous?

2.6 Repeat Exercise 2.5 with the equation $f(x + y) = f(x)f(y)$.

2.7 Find all the solutions of the functional equation (2.4)

1. in the set of all numbers of the form 2^m, where m is an integer,
2. in the set of all numbers of the form 2^r, where r is a rational number,
3. in the set of all numbers of the form a^r, where r is rational and a is a positive constant.

2.8 Repeat Exercise 2.7 with the equation $f(xy) = f(x)f(y)$.

2.9 Find all the solutions of each of the following functional equations in the set of rational numbers:

*1. $f(x + y) + f(x - y) = 2f(x)$;
 2. $f(x + y) + f(x - y) = 2f(x) + 2f(y)$.

2.10 Find all the continuous solutions in R of the equations listed in Exercise 2.9.

*2.11 In the set of all non-negative real numbers, find all the monotone solutions of equation (2.1).

2.12 Find all the continuous solutions of each of the following equations in the set of positive reals:

1. $f(xy) = f(x)f(y)$;
2. $f(x^y) = f(x)$, $y \in R^+$ is variable.

2.13 Find all the continuous solutions of the functional equation

$$f(xy) = \frac{f(x) + f(y)}{x + y}$$

in the interval $(1, \infty)$.

2.14 Find all the real functions that solve, in $(1, \infty)$, the functional equation

$$f(xy) = \frac{f(x) + f(y)}{xy}$$

2.3 OTHER IMPORTANT TYPES OF EQUATIONS

So far, we have been working with functional equations that were not very difficult to solve, although, in connection with continuity for instance, we had to employ Theorem 3. In the present section we give two other typical examples of functional equations, whose solution is a little more difficult. This will give readers an idea of what problems may be encountered when solving functional equations. Also, the given methods may be modified to suit other cases. However, we are still a long way from solving most equations.

One of the difficulties in finding a solution arises from *a priori* restrictions of the domain in which we are seeking the solutions. This is the case if we want to find all the continuous solutions of Jensen's functional equation

$$f\left(\frac{x + y}{2}\right) = \frac{f(x) + f(y)}{2} \tag{2.22}$$

in a closed interval, say $[0, 1]$. Putting $y = 0$ in (2.22), we get

$$f\left(\frac{x}{2}\right) = \frac{1}{2}f(x) + \frac{a}{2}$$

where $a = f(0)$. Thus the given equation yields a new one

$$f\left(\frac{x+y}{2}\right) = \frac{f(x+y)}{2} + \frac{a}{2} = \frac{f(x)}{2} + \frac{f(y)}{2}$$

which reduces to

$$f(x+y) = f(x) + f(y) - a.$$

After the substitution

$$g(x) = f(x) - a \qquad (2.23)$$

we obtain the equation

$$g(x+y) = g(x) + g(y) \qquad (2.24)$$

which, unfortunately, is satisfied only for those $x, y \in [0, 1]$ for which also $x + y \in [0, 1]$. Thus it is an equation of the same type as (2.22). It is called a conditional functional equation because it is only fulfilled for such x and y from the domain of definition that satisfy an additional condition (in our case $x + y \in [0, 1]$).

If the equation (2.22) were satisfied on the whole set R, its solution would be obvious: (2.24) and (2.23) imply $f(x) = bx + a$, where a and b are arbitrary constants. The following theorem and its proof indicate how to solve the original equation.

Theorem 6

Let a continuous function f defined on an interval $[a, b]$ be a solution of (2.22). Then there exist real numbers α, β with $f(x) = \alpha x + \beta$.

The proof is as follows. It is more convenient to solve equation (2.22) in the interval $[0, 1]$. Therefore, instead of f we shall consider the function F defined by

$$F(y) = f((b - a)y + a).$$

It is easy to see that for $y \in [0, 1]$ the number $(b - a)y + a$ falls in the interval $[a, b]$, therefore the function F is well-defined. Moreover, F satisfies Jensen's functional equation (2.22) in the domain $[0, 1]$. In fact,

$$F\left(\frac{x + y}{2}\right) = f\left((b - a)\frac{x + y}{2} + a\right)$$

$$= f\left(\frac{[(b - a)x + a] + [(b - a)y + a]}{2}\right)$$

$$= \frac{f[(b - a)x + a] + f[(b - a)y + a]}{2}$$

$$= \frac{F(x) + F(y)}{2}.$$

Now put $F(0) = c$ and $F(1) = d$. Then

$$F(1/2) = \frac{F(0) + F(1)}{2} = c + \frac{d - c}{2}$$

$$F(1/4) = \frac{F(0) + F(1/2)}{2} = \frac{c + c + (d - c)/2}{2} = c + \frac{d - c}{4}$$

$$F(3/4) = \frac{F(1/2) + F(1)}{2} = \frac{c + (d - c)/2 + d}{2} = c + \frac{3(d - c)}{4}$$

and so on. We will now show that, in general, if x is any number of the form $m/2^n$, where m and n are positive integers satisfying $0 \leqslant m \leqslant 2^n$, then

$$F(x) = c + x(d - c) \qquad (2.25)$$

(the numbers $m/2^n$ are termed dyadic rational numbers). We proceed by induction on n. As we have shown, the assertion is true for $n = 1, 2$. Assume that it holds for $n = k$ and consider two cases:

(a) $\quad x = \dfrac{2m}{2^{k+1}}$ $\qquad$ (b) $\quad x = \dfrac{2m+1}{2^{k+1}}$

In the case (a) we have

$$F\left(\frac{2m}{2^{k+1}}\right) = F\left(\frac{m}{2^k}\right) = c + \frac{m}{2^k}(d-c) = c + \frac{2m}{2^{k+1}}(d-c)$$

and in the case (b)

$$F\left(\frac{2m+1}{2^{k+1}}\right) = F\left(\frac{1}{2}\left(\frac{m}{2^k} + \frac{m+1}{2^k}\right)\right)$$

$$= \frac{1}{2}\left[F\left(\frac{m}{2^k}\right) + F\left(\frac{m+1}{2^k}\right)\right]$$

$$= \frac{1}{2}\left[c + \frac{m}{2^k}(d-c) + c + \frac{m+1}{2^k}(d-c)\right]$$

$$= c + \frac{2m+1}{2^{k+1}}(d-c).$$

Thus (2.25) is satisfied for all elements x of the set of numbers $m/2^n$, $1 \leqslant m \leqslant 2^n$, which is dense in $[0, 1]$ (why?) and hence, from Theorem 3, (2.25) is true for all $x \in [0, 1]$.

Note that restricting the domain in which we solved equation (2.22) did not give us any new solution. This, however, is not always the case. For example, let f be a continuous function defined on R by $f(x) = x + 1$ for $x \in [2, 3]$, $f(x) = x + 2$ for $x \in [4, 6]$ and defined arbitrarily for other x (but so that it is continuous). Readers can easily check that $f(x + y) = f(x) + f(y)$ for all $x, y \in [2, 3]$. Thus f solves the Cauchy functional equation in the domain $[2, 3]$, yet it does not have the form ax for $x \in [2, 3]$.

A type of functional equation which we deal with very often is

$$f(x + y) = F(f(x), f(y)) \qquad (2.26)$$

where f is the unknown function and F is a given function of two variables. Besides the Cauchy equation, the following equations also belong to the said type:

$$f(x + y) = f(x) + f(y) + f(x)f(y) \qquad (2.27)$$

$$f(x + y) = \frac{f(x) + f(y)}{1 - f(x)f(y)} \qquad (2.28)$$

$$f(x + y) = \frac{f(x) + f(y) + 2f(x)f(y)}{1 - f(x)f(y)}. \qquad (2.29)$$

It is not always possible to solve such an equation. If it can be solved, the following method leads to the solution.

From (2.26) we get

$$f(2x) = F(f(x), f(x)) = F_2(f(x))$$

$$f(nx) = F(f((n - 1)x), f(x)) = F_n(f(x))$$

where n is a positive integer and the sequence of functions $F_1, F_2, \ldots$ is defined by

$$F_1(z) = z, \quad F_{n+1}(z) = F(F_n(z), z) \quad \text{for} \quad n = 1, 2, \ldots.$$

The first problem is to find a 'reasonable' rule for F_n, that is, a suitable formula for computing the function F_n. We can only guess here. If we succeed, we continue as follows.

We put $x = 1$ in $f(nx) = F_n(f(x))$ and obtain

$$f(n) = F_n(f(1)) = F_n(c)$$

where $c = f(1)$. Similarly, for $x = m/n$ we get

$$F_n\!\left(f\!\left(\frac{m}{n}\right)\right) = f(m) = F_m(c). \qquad (2.30)$$

If we manage to calculate $f(m/n)$ from the last equality — that is the second difficult point — the worst is already over, because we have a solution defined on the set of positive rational numbers. If we are interested in continuous solutions only, we use Theorem 3. The problem of how to extend the solution to the whole set R depends on the particular equation. The following example will illustrate the above method.

Example 9. Find all the continuous solutions of the functional equation (2.27)

$$f(x + y) = f(x) + f(y) + f(x)f(y)$$

in the domain R.
Solution: We have

$$f(2x) = 2f(x) + f(x)^2 = (1 + f(x))^2 - 1 \qquad (2.31)$$

$$f(3x) = f(2x + x) = 3f(x) + 3f(x)^2 + f(x)^3$$
$$= (1 + f(x))^3 - 1.$$

By induction we infer that

$$f(nx) = (1 + f(x))^n - 1 \qquad (2.32)$$

for all natural n. We have just got through the first difficult point.

Putting $2x = y$ in (2.31), we get

$$f(y) = \left[1 + f\left(\frac{y}{2}\right)\right]^2 - 1 \geqslant -1. \qquad (2.33)$$

After a substitution in (2.30), equation (2.32) yields

$$\left(1 + f\left(\frac{m}{n}\right)\right)^n - 1 = (1 + f(1))^m - 1$$

which simplifies to

$$f\left(\frac{m}{n}\right) = (1 + f(1))^{m/n} - 1. \qquad (2.34)$$

If $f(1) = -1$, then (2.27) implies that

$$f(x) = f(1 + (x - 1)) = f(1) + (1 + f(1))f(x - 1) = -1$$

and hence $f(x) = -1$ for all $x \in R$.

If $f(1) \neq -1$, in view of (2.33) it means that $f(1) > -1$ and, denoting $1 + f(1) = a$ (which is greater than 0), we get from (2.34) that

$$f(r) = a^r - 1 \tag{2.35}$$

for all positive rational numbers r. As we are only seeking continuous solutions, we may use Theorem 3 to obtain the result: (2.35) holds for all non-negative real numbers r. In particular, $f(0) = 0$.

It remains to extend the solution to negative reals. Putting $y = -x$ in (2.27), we get

$$f(0) = f(x) + f(-x) + f(x)f(-x).$$

If x is negative, then

$$f(x) = \frac{-f(-x)}{1 + f(-x)} = \frac{1 - a^{-x}}{a^{-x}} = a^x - 1.$$

We have thus obtained the following result: the only continuous solutions of (2.27) in R are

$$f(x) = -1 \qquad \text{for all} \quad x \in R$$

or

$$f(x) = a^x - 1 \quad \text{for} \quad x \in R$$

where a is any positive constant.

In the exercises below, we introduce further functional equations that can be solved by the methods described in this section. For a more detailed discussion of continuous solutions of functional equations in several variables, we refer readers to the monograph [1].

Exercises

2.15 Let f be a continuous function defined on R and solving the Cauchy functional equation in an interval $[a, b]$. Prove the following assertions:

1. If $[a, b] \cap [2a, 2b] \neq \emptyset$, then $f(x) = cx$ for $x \in [a, b] \cup [2a, 2b]$, c being a suitable constant.
2. If $[a, b] \cap [2a, 2b] = \emptyset$, then $f(x) = cx + d$ for $x \in [a, b]$ and $f(x) = cx + 2d$ for $x \in [2a, 2b]$.

2.16 Find all the continuous solutions in R of

1. equation (2.29),
2. the equation $f(x + y) = \dfrac{f(x) + f(y) - 2f(x)f(y)}{1 - f(x)f(y)}$,

*3. equation (2.28)

2.4 APPLICATIONS OF FUNCTIONAL EQUATIONS IN SEVERAL VARIABLES

Let us return to our fishermen problem of § 2.1. We had to find a non-increasing function f defined on the set of non-negative reals to satisfy the functional equation

$$f(x + y) = f(x)f(y).$$

Such a function was found in Example 5 and Exercise 2.11. The condition demanding that f be non-increasing allows the following possibilities:

1. $f(0) = 1$, $f(t) = 0$ for $t > 0$, or
2. $f(t) = a^t$, where a is a positive constant (and, of course, $a > 1$).

The first case could perhaps correspond to just one fisherman coming and immediately catching a fish. Therefore the problem is solved by the function in the second possibility. How should we interpret it? What does it express? The number na^t is the probable number of anglers who are still near the pool at the time t. Evidently, the higher the number of fishermen, n, at the beginning, the better the above function describes the situation.

The example involving fishing may have seemed artificially constructed, but the same situation occurs very often in practical problems. Therefore we give some of the more relevant practical examples.

Example 10. Nuclei of radioactive elements spontaneously split, giving rise to other elements. Let us try to find how this fission depends on time. Consider an initial amount m (in grams) of a radioactive element and let $mf(t)$ denote its quantity at the time t. Analogously to the example about fishermen, we make the assumption (which has been verified experimentally) that, in equal time intervals, the ratio between the quantity of the element at the beginning of the interval and that at its end is the same. Consider two equally long time intervals $[0, h]$ and $[t, t + h]$. Our assumption gives

$$m/(mf(h)) = (mf(t))/(mf(t + h))$$

that is

$$f(t + h) = f(t)f(h).$$

Hence, the amount of the element at the time t will be ma^t, where a is a positive constant ($a < 1$) depending only on the element in question (the fission rate differs for different chemical elements). In this case, the function ma^t will exactly reflect reality, as the number of atoms in a sample is very big.

Example 11. The amplitude of a swinging pendulum (say, a sufficiently heavy weight suspended on a long thread) decreases due to air resistance and, when the amplitude is small enough, the pendulum oscillates in what is called damped harmonic motion. We wish to find the dependence of the amplitude on time. If the pendulum is swinging over a small arc, the amplitude will decrease in equal proportions in equal intervals of time. This reasoning leads again to equation (2.1).

Example 12. A rocket flying in cosmic space has initial mass m_0. After activating its engines for some time, its mass decreases to m_1 (due to burning of fuel) and the rocket attains a certain velocity. We are interested to find out what the velocity increase, δ, depends on.

Solution: The central factor is the ratio m_0/m_1. Denote by $\delta(m_0/m_1)$ the speed increment corresponding to the change of mass in the ratio m_0/m_1. When the rocket passes from the state m_0 through m_1 to a state m_2, its velocity increment can be expressed in two ways: either by $\delta(m_0/m_2)$ or as the sum $\delta(m_0/m_1) + \delta(m_1/m_2)$ (assuming that the rocket does not change its direction, we consider the absolute values of speed increments). Thus we obtain the equality

$$\delta(m_0/m_2) = \delta(m_0/m_1) + \delta(m_1/m_2).$$

Putting $m_0/m_1 = x$ and $m_1/m_2 = y$, we get the following functional equation for δ:

$$\delta(xy) = \delta(x) + \delta(y).$$

We are seeking its solution in the set of real numbers that are greater than or equal to 1. If, moreover, the function δ is assumed monotone (with a growing ratio m_0/m_1, the increment $\delta(m_0/m_1)$ grows as well) or continuous (the increase of velocity has no jumps), we obtain $\delta(x) = c \ln x$ for all $x \geqslant 1$, where c is a constant (depending on the construction of the rocket, the fuel used, etc). (See Exercise 2.8.)

Example 13. Let us return once more to the anglers' problem. Suppose that they do not come to the pool all at the same time, but one after another in such a way that in equal time intervals the same number of them arrive. We wish to know how the number of fishermen present at the pool depends on time.

Solution: We measure the time from the moment when the first angler starts fishing. The number of fishermen present at the pool at a time t will be denoted by $h(t)$. We seek the function h. The number of anglers fishing at the time $t + s$ may be expressed in two ways: either by $h(t + s)$ or as follows. There are $h(t)$ fishermen present at the time t and, by the time $t + s$, only $h(t) a^s$ of those $h(t)$ men will have stayed there (a is the constant from the first problem concerning the anglers) and $h(s)$ other fishermen will have come. We obtain the functional equation

$$h(t + s) = h(t) a^s + h(s) \qquad (2.36)$$

which can be solved as follows. As well as equation (2.36), it is true that

$$h(t + s) = h(s + t) = h(s) a^t + h(t)$$

and the last two equations yield

$$h(t) a^s + h(s) = h(s) a^t + h(t).$$

Putting $s = 1$ in the last equality we get

$$h(t) = h(1) \frac{a^t - 1}{a - 1}$$

for all $t \geqslant 0$. Sketch the graph of the function h. The number of fishermen present at the pool will increase first, and later it will stabilise (calculate the limit of $h(t)$ as $t \to \infty$, you must make use of the fact that $a < 1$).

The following exercises deal with situations similar to those described in the above examples. All of them show, besides other things, the importance of the functions a^x, or e^x, and $\ln x$, which very often model real processes. For other, less trivial, cases in which functional equations are applied, we refer the readers to the literature listed in the bibliography at the end of this book.

Exercises

2.17 Water absorbs light. It is well known that very deep in the sea it is dark. Find out how the intensity of a light ray coming from sea-level depends on the depth of water. (Hint: the intensity of the light passing through equally thick layers of water is decreased in the same proportion.)

2.18 A sum of money deposited in a bank increases due to interest. Examine how it depends on time (assuming that money was deposited only once).

2.19 Find out how the amount of money in the bank increases with time if we assume regular savings. (Hint: review Example 13.)

2.20 Bombardment of certain elements by neutrons (in an atomic reactor, for example) gives rise to radioactive isotopes which are unstable and prone to fission. Let a sample containing a certain amount of such an element be put into a reactor. Suppose that only a very small part of the element is changed into the radioactive isotope. Therefore, the isotope is formed in the sample at a constant rate, that is, the same quantity of the isotope is formed in equally long intervals of time. Determine the total amount $h(t)$ of the isotope in the sample at time t. Investigate the function h and explain why, after some time, the amount of the isotope in the sample will remain approximately constant.

*2.21 Fission of a radioactive element X gives rise to a

radioactive element Y, which in its turn produces a stable element Z. At the time $t = 0$, we begin to observe a certain amount of pure element X. After some time, we obtain a mixture of all three elements. Determine the proportion of individual elements in the mixture.

*2.22 How does the answer to the preceding problem change if the element X is produced in the reactor at a constant rate from a stable element, as in Exercise 2.20?

2.5 DISCONTINUOUS SOLUTION OF THE CAUCHY FUNCTIONAL EQUATION

In this section we will show that discontinuous solutions of the Cauchy functional equation in R also exist, that is solutions other than $f(x) = ax$ with a constant. We already know something about such a discontinuous function f (if, of course, it exists): on no interval can it be bounded. This is implied by Theorem 4.

This means that such a function f has the following property: if n is any positive integer, then the set A_n of all those points x at which $f(x) > n$ is dense in R (that is, every interval contains some point of A_n). This property indicates that it will not be easy to draw a graph of such a function. The following theorem gives some idea of the discontinuous solutions of the Cauchy equation.

Theorem 7

Let f be a discontinuous function solving the Cauchy functional equation

$$f(x + y) = f(x) + f(y)$$

in the domain R. Let $I = (a, b)$, with $a < b$, be any interval. Then the set $f(I)$ of all the numbers $f(x)$ with $x \in I$ is dense in R.

The proof of this is as follows. Suppose that the assertion of the theorem is false. That means that there is an interval $J = (c, d)$ such that for all $x \in I$ we have $f(x) \notin J$, that is, for all $x \in I$ either

$$f(x) < c \qquad \text{or} \qquad f(x) > d. \qquad (2.37)$$

Choose a number $\delta > 0$ to satisfy

$$c + \delta < d - \delta.$$

We may already suppose that the interval I is so small that it is mapped by the function $\varphi(x) = xf(1)$ onto an interval shorter than δ (otherwise we replace I by a smaller interval $I' \subset I$).

Let us now examine the function

$$g(x) = f(x) - xf(1).$$

Using (2.37), we easily check that for all $x \in I$ we have

$$g(x) < c + \delta \qquad \text{or} \qquad g(x) > d - \delta.$$

Hence,

$$g(x) \notin J' = (c + \delta, d - \delta) \qquad (2.38)$$

for all $x \in I$. We shall produce a contradiction by showing that this is not true.

To begin with, observe that for some $x_0 \in I$ we have $f(x_0) \neq x_0 f(1)$, that is $g(x_0) \neq 0$, otherwise the function f would be bounded on I and hence, from Theorem 4, it would be continuous on R. Choose a rational number r with $rg(x_0) \in J'$ and a rational number s such that $y = rx_0 + s \in I$. Then

$$g(y) = f(rx_0 + s) - (rx_0 + s)f(1)$$
$$= rf(x_0) + sf(1) - rx_0 f(1) - sf(1)$$
$$= rf(x_0) - rx_0 f(1) = rg(x_0)$$

(we have employed equation (2.13) and Theorem 1), which means that $g(y) \in J'$ and this contradicts (2.38). The theorem is proved.

The above theorem has caused many mathematicians to believe that every solution of the Cauchy functional equation is necessarily continuous. It was only thanks to the development of the set theory at the end of the 19th and the beginning of the 20th centuries that the existence of a discontinuous solution could be proved. The first to prove this was the German mathematician G Hamel in 1905. We give his argument here. The proof is based on the existence of a special set of real numbers, today called a Hamel basis of real numbers. It is defined as follows.

Definition. A subset H of the set R of real numbers is called a Hamel basis if it has the following property. Every real number $x \neq 0$ can be uniquely expressed up to the order of the summands in the form

$$x = r_1 h_1 + \ldots + r_n h_n \tag{2.39}$$

where n is a suitable positive integer (depending on x), h_i are pairwise distinct elements of H and r_i are suitable non-zero rational numbers.

Every definition should be illustrated by an example. Unfortunately, this is not possible in this case. No concrete example of such a basis is known, we only know that it exists. More precisely, its existence follows from a non-trivial axiom of the set theory, namely the Axiom of Choice. That is why we do not present here the proof of its existence. For our purpose it is sufficient to accept the existence of a Hamel basis as a fact. Readers wishing to know more about the topic may consult the monographs [8], [17] or [27].

The following theorem illustrates how it is possible to

construct solutions of the Cauchy equation using a Hamel basis.

Theorem 8

Let H be a Hamel basis and let g be any mapping of H into R. Then the function f which assigns to every $x \in R$ of the form (2.39) the value

$$f(x) = r_1 g(h_1) + \ldots + r_n g(h_n)$$

is a solution of the Cauchy functional equation.

To prove this, observe first that for every $x \in R$ there is exactly one corresponding number $f(x)$; hence f is a function defined on R. This is where we make use of the uniqueness of the expression (2.39).

Now let x and y be any real numbers. If we allow the rational coefficients in the development of the equation of the type (2.39) to be equal to zero, then both x and y can be expressed using the same elements $h_1, \ldots, h_n$ of H, where n is a suitable positive integer. Thus x may be expressed by (2.39) and y in the form

$$y = s_1 h_1 + \ldots + s_n h_n$$

where r_i and s_i are rational numbers, not necessarily different from zero. Then

$$
\begin{aligned}
f(x) + f(y) &= f(r_1 h_1 + \ldots + r_n h_n) + f(s_1 h_1 + \ldots + s_n h_n) \\
&= r_1 g(h_1) + \ldots + r_n g(h_n) + s_1 g(h_1) + \ldots + s_n g(h_n) \\
&= (r_1 + s_1) g(h_1) + \ldots + (r_n + s_n) g(h_n) \\
&= f((r_1 + s_1) h_1 + \ldots + (r_n + s_n) h_n) \\
&= f(x + y)
\end{aligned}
$$

hence f solves the Cauchy equation and the theorem is proved.

Now, finally, let us consider a discontinuous solution.

Example 14. Let H be a Hamel basis and let $h_0 \in H$ be fixed. Define a function f by means of Theorem 8 in such a way that g has the following properties:

$$g(h_0) = 1 \qquad \text{and} \qquad g(h) = 0$$

for $h \neq h_0$, $h \in H$. It is easy to verify that if the element h_0 multiplied by r_0 occurs in the development of x using H, then $f(x) = r_0$, otherwise $f(x) = 0$. Evidently, f is not of the form $f(x) = ax$ because it assumes the value zero at infinitely many points. More precisely, $f(sh) = 0$ whenever $h \in H$, $h \neq h_0$ and s is any rational number.

Further examples of discontinuous solutions of the Cauchy equation are given in the exercises below.

Exercises

2.23 Let X be the set $Q + Q2^{1/2}$ of all numbers of the form $r + s2^{1/2}$, where r and s are rational numbers. Show that $H = \{1, 2^{1/2}\}$ is something like a Hamel basis of the set X (see also Exercise 2.5, part 2).

2.24 Let $Y = Q + Q2^{1/2} + Q3^{1/2}$.

1. Solve the analogous problem as in the preceding exercise, for Y.
2. Find all the solutions of the Cauchy equation in the domain Y.

2.25 Show that all the solutions of the Cauchy equation, including the continuous ones, can be obtained by the method indicated in Theorem 8.

2.26 Prove that each of the equations

1. $f(x + y) = f(x)f(y)$

2. $f(xy) = f(x)f(y)$
3. $f(xy) = f(x) + f(y)$

has a discontinuous solution. (Hint: use a discontinuous solution of the Cauchy equation.)

*2.6 CONCLUDING REMARKS

In the preceding sections we have dealt with some elementary results only concerning functional equations in several variables, and it was intended as an introduction to their study. Readers can find further useful information in monographs [1], [8] and [17], already mentioned above (readers will be surprised to see how much mathematicians already know about functional equations). Many important results are only a few years old; there are still many questions unanswered. Two interesting problems are mentioned below. Both of them are related to the Cauchy functional equation.

In §2.2 we have shown that every solution of the Cauchy equation which is upper bounded on some interval is continuous. A natural problem arises—are there other sets, besides intervals, with the same property? Can we describe them?

The problem can be exactly formulated as follows. Let $\mathscr{B}$ be the family of all subsets of R with the property that if $A \in \mathscr{B}$ and if f is a solution in R of the Cauchy equation upper bounded on A, then f is a continuous solution (in R). The question is, which sets belong to $\mathscr{B}$? Theorem 4 says that $\mathscr{B}$ contains all intervals and we infer from Example 14 that no Hamel basis is in $\mathscr{B}$. It can also be proved (see Exercise 2.27 below) that if f is a solution of the Cauchy equation upper bounded on a set $T \subset R$, then f is upper bounded also on the set $T + T$ of all numbers $x + y$ with x, $y \in T$. Therefore, if $T + T$ contains an interval, then $T \in \mathscr{B}$.

In this way it can be shown that $\mathscr{B}$ contains any set of positive Lebesgue measure and some other sets.

Similarly, we may introduce a family $\mathscr{C}$ such that $T \in \mathscr{C}$ if, and only if, every solution of the Cauchy equation bounded on T (i.e. both upper and lower bounded) is continuous. It is easy to see that $\mathscr{B} \subset \mathscr{C}$. It can be checked (see Exercise 2.28 below) that $\mathscr{B} \neq \mathscr{C}$.

It was only recently that the families $\mathscr{B}$ and $\mathscr{C}$ were characterised (see [28] and [29]; simplified proofs, written more methodically, can be found in [2], [8] and [17]). In the case of $\mathscr{C}$, the characterisation is rather simple: $T \in \mathscr{C}$ if, and only if, the Q-convex hull of $T - T = \{x - y; \; x, y \in T\}$ includes an interval. Here the Q-convex hull of a set A means the set of all numbers of the form $a_1 x_1 + \dots + a_n x_n$, where $x_i \in A$, $a_i \in Q$, $a_i > 0$, $a_1 + \dots + a_n = 1$ and n is any positive integer.

Let us mention now a similar, but so far unsolved, problem. If f is a solution of the Cauchy equation in R which is continuous on an interval I, then f is continuous on the whole of R. This assertion is a simple consequence of Theorem 4. Let us now ask which other sets besides intervals have the analogous property? More precisely, let $\mathscr{D}$ be the family of all $T \subset R$ with the property that whenever f is a solution of the Cauchy equation in R and the restriction $g = f|T$ of f to T is a continuous function $T \to R$, then f is continuous on R. Which sets belong to $\mathscr{D}$? It is well known that every set of the second category with the Baire property is in $\mathscr{D}$. Yet no characterisation of the family $\mathscr{D}$ is known and the author believes that such a characterisation would not be very simple. For more information about the problem, we refer readers to [14], where they can also find some older citations.

Several open problems concerning functional equations are presented in the monograph [17].

Exercises

2.27 Prove that if f is a solution in R of the Cauchy equation which is upper bounded (or bounded, respectively) on a set $T \subset R$, then f is upper bounded (or bounded, respectively) also on the set $T + T = \{x + y;\ x, y \in T\}$.

2.28 Let H be a Hamel basis. Let H^* be the set of all numbers of the form $|h|$, where $h \in H$. Then H^* is a Hamel basis as well.

1. Prove this.
*2. Prove that if T is the set of all numbers $a_1 h_1 + \ldots + a_n h_n$ with $h_i \in H^*$, $a_i \in Q$, $a_i < 0$ and n is any positive integer, then $T \in \mathscr{C}$ and $T \notin \mathscr{B}$.

3. ITERATIONS

3.1 INTRODUCTION

Did you ever try to solve the equation

$$\sin x = x \qquad \text{or} \qquad \cos x = x$$

(with x measured in radians)? For the former equation, we can easily check that $x = 0$ is a solution, and a short reasoning verifies also that it is the only solution. It is not so easy to find a solution of the latter equation, especially just using a pencil and a sheet of paper, the classical tools of a mathematician. Well, let us try to solve it with a calculator. Choose any number $x = x_0$, calculate $x_1 = \cos x_0$, then $x_2 = \cos x_1 = \cos(\cos x_0)$, and so on. We obtain a sequence of numbers $x_0, x_1, x_2, \dots$ which converges to $0.73908 \dots$ The exact value depends on how many decimal places the calculator works with. The number obtained is an approximate solution of the given equation.

If we apply a similar method to the first equation, the resulting sequence will stabilise near the number zero; the more exactly the calculator works, the nearer the result will be to zero. Thus, we again get an approximate solution of our equation. For both equations, the above method works reliably. With whatever number we start, we always get the same result. This also implies that each of the said equations has exactly one solution.

We know that there exist solutions of the equations

$$\tan x = x \qquad \cot x = x$$

and that there is an infinite number of them; to visualise this just sketch on one diagram the graphs of $f(x) = \tan x$ and $g(x) = x$.

The points where the two graphs intersect represent the solutions of the equation $\tan x = x$ (see figure 3.1). These equations, however, cannot be solved by the same method which proved efficient in the case of $\sin x$ and $\cos x$, and which is termed the iteration method. In many cases, however, the method of iterations is the only means for solving

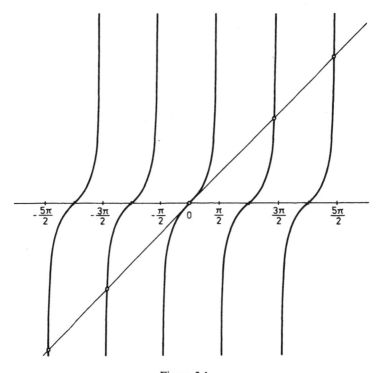

Figure 3.1

equations, computing the values of certain functions etc. What its applicability depends on will be discussed in the next section. Now we introduce another example.

Many situations in various domains of science can be, at least approximately, mathematically modelled in a very simple way. As an illustration, we take an example from biology. Suppose that in a given environment there are certain animals, say, rabbits, and that we are interested to know how their number varies with time. Suppose that their number is measured once a year, every spring. At the beginning we have x_0 individuals, after a year x_1, a year later x_2 etc. Under certain, partly idealised conditions, the number of individuals in the year $n + 1$ depends only on x_n. This may be expressed by

$$x_{n+1} = f(x_n), \quad n = 0, 1, 2, \ldots \tag{3.1}$$

where f is an unknown function. We obtain a sequence, constructed analogously to the case of solving the equation $\cos x = x$. A similar model can be used in other sciences, such as epidemiology, with x_n denoting the percentage of infected persons at the time n, genetics, economics, meteorology, sociology, etc. We are not concerned now with the problem of how to find the function f. An appropriate type of function is usually proposed on the basis of measurement or observation.

In practice, x_n usually cannot grow indefinitely; this often implies that for small x_n we have $x_{n+1} > x_n$, but then for large x_n we get $x_{n+1} < x_n$. Therefore, assume that $0 \leqslant x_n \leqslant a$. Then the function f maps the interval $I = [0, a]$ into I as illustrated, for instance, in figure 3.2. Readers can easily check that for $0 < x < c$ we have $f(x) > x$, for $x > c$ we have $f(x) < x$, and $x = c$ or $x = 0$ gives $f(x) = x$.

A graph of an experimental function f is approximated by an explicitly defined function containing parameters that

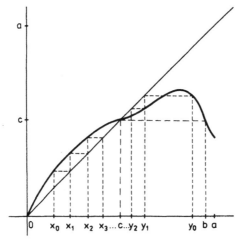

Figure 3.2

are subsequently determined by means of statistical methods (which we shall not do in this book). Let us list some functions that have actually been used for that purpose:

$$Ax(1 - x) \qquad x\,e^{A(1 - Bx)}$$
$$x[1 + A(1 - Bx)] \qquad Ax(1 + e^{-B(1 - Cx)})^{-1}$$
$$Ax(1 + Bx)^{-C} \qquad Ax(1 + Bx^{C})^{-1}$$

and others, where A, B, C are real, constant, parameters.

The following problems arise in practice:

1. How reliable is the model?

2. Having already chosen a certain function f, we have to predict the possible evolution of the population over a rather long time interval (which mathematically means to describe the properties of the sequence $\{x_n\}_{n=0}^{\infty}$ given by (3.1) for various initial values x_0).

3. Will the answer to the problem listed under 2 change if the chosen function f is replaced by a function $\tilde{f}$, which

differs only a little from f (that is $|f(x) - \tilde{f}(x)|$ is very small for all $x \in I$)?

It is beyond both the ability of the author and the aim of this book to give a satisfactory answer to the first of the listed questions. Therefore, in the following we shall concentrate on the second and also, partially, the third problem.

Let us introduce another useful notation. Given a function f mapping an interval I into I and a positive integer n, we shall denote by f^n the composite function, mapping I into I again, that may be defined recursively by

$$f^1(x) = f(x), \qquad f^{n+1}(x) = f(f^n(x)) \quad \text{for} \quad n = 1, 2, \ldots.$$

The function f^n is termed the nth iterate of f. Observe that if f is a continuous function, then all its iterates are continuous as well. Instead of (3.1), we may write

$$x_n = f^n(x_0), \qquad n = 1, 2, \ldots.$$

The sequence $\{x_n\}_{n=0}^{\infty}$ given by the above rule is said to be generated by f and x_0.

A point $x_0 \in I$ with $f(x_0) = x_0$ is called a fixed point of f. The sequence generated by a function f and a fixed point of f is stationary, which means that all its terms are the same. In our model this represents a stationary regime: the number of individuals is the same in all generations.

In many cases, we can deduce some properties of sequences generated by certain points from the graph of the function. In fact, given a point x_0, it is easy to find its image $x_1 = f(x_0)$, then the image $x_2 = f(x_1)$ of x_1 etc. For instance, it is evident from figure 3.2 that the sequence generated by x_0 converges to the fixed point c. A more detailed study of the graph reveals other information. If x_0 is any point in $I_1 = (0, c)$, then the sequence generated by x_0 is increasing and converges to c. The point b on the x axis is also important; it is the only point distinct from c with the property $f(b) = c$. The sequence generated by any point $y_0 \in I_2 = (c, b)$ is decreasing and also converges to c. Finally, if $z_0 \in I_3 = (b, a)$, then $f(z_0) \in I_1$ and, hence, $z_1, z_2, z_3, \ldots$ is an increasing sequence converging to c. The remaining points 0 and c are fixed points, and they generate stationary sequences. Thus the function shown in figure 3.2 has the particular property that all the points of I generate convergent sequences, the interpretation being that the evolution of the population will stabilise after some time, no matter what the original state x_0 was.

As we shall see later, not all functions generate such nice sequences; recall the functions $\tan x$ and $\cot x$ that we have already mentioned.

3.2 FIXED POINTS

Let us review our example of solving the initial equation $\sin x = x$. Figure 3.3 shows the sequence $x_0 = 1$, $x_1 = \sin x_0, \ldots, x_{n+1} = \sin x_n$. In fact, for any $x > 0$ we have $\sin x < x$, and so our sequence will be decreasing as

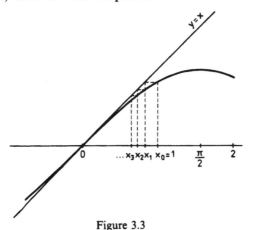

Figure 3.3

$x_{n+1} = \sin x_n < x_n$. Similarly, choosing $x_0 < 0$, we would get an increasing sequence.

The corresponding sequence for the equation $\cos x = x$, again with $x_0 = 1$, is sketched in figure 3.4. This sequence is not monotone, but it is convergent. More exactly, it can be decomposed into two monotone sequences: $x_0 = 1 > x_2 > x_4 > \ldots$ is a decreasing sequence and $x_1 < x_3 < x_5 < \ldots$ is an increasing one. Both have the same limit, $\alpha = 0.73908 \ldots$.

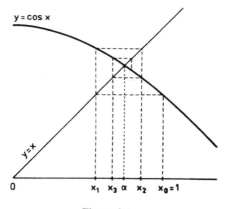

Figure 3.4

Figure 3.5 depicts an analogous process for an equation $f(x) = x$, where the function f is given by its graph. However, if we choose the initial point x_0 arbitrarily close to α but so that $x_0 \neq \alpha$, the sequence will run away from α in both directions. A similar situation occurs in the case of both $\tan x$ and $\cot x$ in the neighbourhood of any of their fixed points.

Figures 3.3 to 3.5 illustrate different types of fixed points. The fixed point c in figure 3.2 is of the same type as that in figure 3.3. Evidently, the answer to the question of whether a fixed point of a function f (or a solution of an

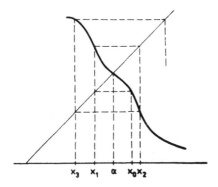

Figure 3.5

equation $f(x) = x$) can be found by the method of iterations depends on the behaviour of f in the neighbourhood of that fixed point. The following theorems illustrate the dependence. First we introduce some basic ideas we shall need.

Definition 1. A fixed point α of a function f defined on an interval I is said to be

1. attractive if there exists a neighbourhood V of α (i.e. an open interval containing α) such that for any $x_0 \in V$ the sequence of iterates $\{f^n(x_0)\}_{n=1}^{\infty}$ converges to α;

2. repulsive if there exists a neighbourhood V of α such that for every $x_0 \in V$ with $x_0 \neq \alpha$ we have $f^n(x_0) \notin V$ for some n.

In other words, if α is an attractive fixed point, then the sequence generated by any point x_0 from some neighbourhood V of α (where $x_0 \neq \alpha$) converges to α. If α is a repulsive fixed point, then the sequence will run out of V. We see that the fixed points in figures 3.2, 3.3 and 3.4 are attractive, while the fixed point α in figure 3.5, as well as all the fixed points of the functions $\tan x$ and $\cot x$, are examples of repulsive fixed points.

Attractive and repulsive fixed points are rather an

exception among the fixed points, as far as their occurrence is concerned. For instance, every point $x \in R$ is a fixed point of the identity function $f(x) = x$ defined on R, but each of them is easily verified to be neither attractive nor repulsive. This is implied also by the following theorem.

Theorem 1

Let f be a continuous mapping of an interval I into I. Every attractive or repulsive fixed point α of f is isolated.

This means that some neighbourhood U of α does not contain any other fixed point.

The proof is as follows. If every neighbourhood U of α contains a fixed point $\beta \neq \alpha$, then the sequence generated by β is stationary and will neither converge to α nor run out of U. Thus α is neither an attractive nor a repulsive fixed point.

There are functions, even continuous ones, that have no fixed points. Such functions as e^x or $\log x$ may serve as an example. Thus the equation $e^x = x$ has no solution in R. Later, we shall find the next theorem very useful. It guarantees the existence of fixed points in some cases.

Theorem 2

Let f be a continuous function on an interval I and let the inequalities $f(x) > x$ and $f(y) < y$ be satisfied for some $x, y \in I$. Then f has a fixed point between x and y.

We can prove this as follows. Consider the function $g(t) = f(t) - t$. Evidently, $g(x) > 0$ and $g(y) < 0$. Since g is continuous, it assumes the value zero at some point α between x and y. Then, of course, $f(\alpha) = \alpha$.

Corollary. Let f be a continuous mapping defined on a closed interval I. If $f(I) \subset I$, then f has a fixed point in I.

This can be proved by letting $I = [a, b]$ and assuming that neither a nor b is a fixed point. Then $f(a) > a$ and $f(b) < b$, and it suffices to apply Theorem 2.

The next theorem gives conditions for a fixed point to be attractive or repulsive.

Theorem 3

Let f be a continuous function mapping an interval I into I and let α be its fixed point.

1. If all $x \neq \alpha$ in some neighbourhood U of α satisfy

$$\left| \frac{f(x) - f(a)}{x - a} \right| < 1 \tag{3.2}$$

then α is an attractive fixed point.

2. If for all $x \neq \alpha$ from some neighbourhood U of α we have

$$\left| \frac{f(x) - f(a)}{x - a} \right| > 1 \tag{3.3}$$

then α is a repulsive fixed point.

The condition (3.2) means that the graph of f in a neighbourhood of α lies in the area marked by shading in figure 3.6a, while (3.3) says that the graph lies in the area shaded in figure 3.6b.

The proof is as follows. Let $x_0 \in U$ be a number with $x_0 \neq \alpha$. Denote $x_{n+1} = f^n(x_0)$ for $n = 1, 2, \ldots$. Putting $x = x_n$ in (3.2) we get

$$|f(x_n) - f(a)| < |x_n - a|$$

and owing to $f(a) = a$ and $f(x_n) = x_{n+1}$ we have

$$|x_{n+1} - a| < |x_n - a|.$$

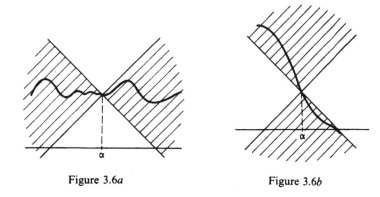

Figure 3.6a   Figure 3.6b

We see that the sequence $a_n = |x_n - \alpha|$ is decreasing and lower bounded (by zero, for instance), hence it converges to some limit a.

It is sufficient to show that $a = 0$. Suppose, on the contrary, that $a > 0$. From (3.2) it follows that $f(\alpha - a) \in (\alpha - a, \alpha + a) = J$. Since f is a continuous function, there exists a neighbourhood U_- of $\alpha - a$ with $f(U_-) \subset J$. Analogously, there is a neighbourhood U_+ of $\alpha + a$ satisfying $f(U_+) \subset J$. As $\lim_{n \to \infty} |x_n - \alpha| = a$, for some n we have $x_n \in U_-$ or $x_n \in U_+$. Then, of course, $x_{n+1} = f(x_n) \in J$, hence $|x_{n+1} - \alpha| < a$, which is impossible. The obtained contradiction proves that $a = 0$ and x_n converges to α. The proof of part 1 is complete. The assertion in part 2 can be proved analogously.

If f has a derivative at a fixed point α, the following theorem allows us to determine the character of the fixed point in some cases.

Theorem 4

Let a continuous function f, mapping an interval I into I, have a derivative at a fixed point $\alpha \in I$. Then

1. *if $|f'(a)| < 1$, a is an attractive fixed point*;
2. *if $|f'(a)| > 1$, a is a repulsive fixed point*.

To prove this, let us start by putting $|f'(a)| < 1$. It follows from

$$f'(a) = \lim_{x \to a} \frac{f(x) - f(a)}{x - a}$$

that (3.2) is true for all $x \neq a$ sufficiently close to a, and it suffices to use Theorem 3. The proof of part 2 is analogous.

To conclude this section, we shall state one more theorem. Later, we shall appreciate its usefulness.

Theorem 5

Let f be a continuous function, mapping an interval I into I, and suppose that for some $x_0 \in I$ the sequence $\{f^n(x_0)\}_{n=1}^\infty$ converges to some point a. Then a is a fixed point of f.

The proof is as follows. Let $\varepsilon > 0$. Since the function f is continuous at a, there exists $\delta > 0$ such that for any $y \in (a - \delta, a + \delta)$ we have

$$|f(y) - f(a)| < \varepsilon. \tag{3.4}$$

We may assume that $\delta < \varepsilon$ (otherwise we take a smaller δ). Since $f^n(x_0)$ converges to a, for all sufficiently large n we have

$$|f^n(x_0) - a| < \delta < \varepsilon \tag{3.5}$$

and, hence, from (3.4)

$$|f(f^n(x_0)) - f(a)| = |f^{n+1}(x_0) - f(a)| < \varepsilon. \tag{3.6}$$

Now (3.5) and (3.6) imply that

$$|f(a) - a| \leqslant |f(a) - f^{n+1}(x_0)| + |f^{n+1}(x_0) - a|$$
$$< \varepsilon + \varepsilon = 2\varepsilon$$

for all sufficiently large n. We have thus shown that $|f(\alpha) - \alpha|$ is less than any positive number, hence $f(\alpha) - \alpha = 0$ and α is a fixed point. The theorem is proved.

Further examples of fixed points with various properties are included in the exercises.

Exercises

3.1 Decide which of the fixed points of the following functions are attractive and which are repulsive.

1. $f(x) = 2 \sin x$.
2. $f(x) = 2 \cos x$.
3. $f(x) = x^2 \sin(1/x)$ for $x \neq 0$ and $f(0) = 0$.
4. $f(x) = 1 - x^2$.
5. $f(x) = 1 - |1 - 2x|$ for $x \in [0, 1]$.
6. $g(x) = f^2(x) = f(f(x))$, where f is the function defined in part 4.

3.2 Find a polynomial function of degree 2 having two fixed points, both of them being computable by the method of iterations starting in some suitable neighbourhood of the fixed point.

3.3 Prove that if I is a closed interval and $f: I \to R$ is a continuous function with $f(I) \supset I$, then f has a fixed point in I.

3.4 Let $x_0 = 1$ and $x_{n+1} = x_n + x_n^{-2}$ for $n = 0, 1, 2, \ldots$.

1. Is the sequence $\{x_n\}_{n=1}^{\infty}$ bounded?
2. Show that $x_{9000} > 30$.

3.5 How many solutions has the following system of equations?

$$x_2 = 1 - x_1^2 \qquad x_4 = 1 - x_3^2$$
$$x_3 = 1 - x_2^2 \qquad x_1 = 1 - x_4^2$$

(Hint: x_1 is necessarily a fixed point of a certain function.)

3.6 A fixed point of a continuous function f may be attractive even if the condition (3.2) is not met. Analogously, a fixed point α may be repulsive without (3.3) being fulfilled. Give some examples.

3.7 Theorem 4 does not mention the case $f'(\alpha) = \pm 1$.

1. Show that, in that case, α may be either attractive or repulsive (of course, not at the same time).
*2. Show that yet another situation may occur, even if α is an isolated fixed point.

3.8 Try to sketch the graph of a continuous function in a neighbourhood of its isolated fixed point α so that it has the following property.

1. Every neighbourhood V of α contains two points $x_0 \neq \alpha$ and y_0 such that the sequence generated by x_0 converges to α and the sequence generated by y_0 runs out of V.
*2. Can you put down an exact definition of such a function? (Hint: combine the two functions from Exercise 3.6.)

3.9 Definition 1 provides only a rough classification of fixed points. Classify in more detail the isolated fixed points of continuous functions. (Hint: complete figure 3.6 by drawing the straight lines $x = \alpha$ and $y = \alpha$ and examine in which area lies the graph on the left and on the right of α.)

3.10 A function $f: I \rightarrow I$, where I is an interval, is said to be contractive if there exists a positive constant $k < 1$ such that for all $x, y \in I$ we have

$$|f(x) - f(y)| \leqslant k|x - y|.$$

1. Show that a contractive function on an interval is continuous.

2. Prove that if I is an interval of the type $[a, b]$, $(-\infty, a]$, $[b, \infty)$ or $(-\infty, \infty)$, then every contractive function has exactly one fixed point in I.

3. Show that the assertion in part 2 does not hold for an open bounded interval I.

3.3 CYCLES

In a mathematical model, such as that of a population growth presented in the introduction to this chapter, a fixed point represents a stationary regime. This means that whenever the population attains the value $x_n = \alpha$, where α is a fixed point, it will not change any more. The same (from the practical point of view) is true if the sequence $\{x_n\}_{n=0}^{\infty}$ converges (to a fixed point, from Theorem 5). However, a stationary regime may also be represented by some non-convergent sequences, for instance the periodic sequence 1, 2, 3, 1, 2, 3, ... with period 3. Therefore, in what follows we shall examine in more detail the functions that generate periodic sequences. The fundamental concepts here are those of a cycle and of its order. Here is the definition.

Definition 2. Let f be a function mapping an interval I into I. We say that a point $x_0 \in I$ is a periodic point of order n of the function f, or that x_0 generates a cycle of order n, if $f^n(x_0) = x_0$ and $f^i(x_0) \neq x_0$ for $i = 1, 2, ..., n - 1$.

Evidently, the sequence generated by such a point x_0 is periodic with (the basic) period n. It is not difficult to find an example of a function generating only periodic sequences.

Example 1. Consider the function $f(x) = -x$, defined on the whole set R. It has a unique fixed point $\alpha = 0$ (which is in fact a periodic point of order 1) and all the other points are periodic points of order 2; each of them generates a periodic sequence $x_0, -x_0, x_0, -x_0, \ldots$ with period 2.

Also, it is easy to find a function f (in general, not continuous) having only cycles of given orders.

Example 2. Define $f: R \to R$ by $f(0) = 1$, $f(1) = 2$, $f(2) = 0$ and $f(x) = 0$ for other x. This function has exactly one 3-cycle, consisting of points 0, 1, 2, and has no other cycle (nor a fixed point). However, our function has the disadvantage of not being continuous.

If we tried to find a continuous function on R having exactly one cycle of order 3 and no other cycles, we would never succeed. As we shall see later, the existence of cycles of continuous functions obeys certain laws. For example, a continuous function having a 3-cycle must have cycles of all orders including fixed points. That is why we shall consider only continuous functions.

Let us now present a more extensive example of a function depending on a parameter that will show how cycles of various orders are produced by changing the parameter.

Example 3. Let us examine in more detail the function $f_A(x) \cdot Ax(1 - x)$ mentioned in §3.1. As we stated earlier, this function is used for modelling certain phenomena, in biology for instance. It is easy to see that f_A is a continuous function from $[0, 1]$ to $[0, 1]$ if $A \in [0, 4]$. Figure 3.7 shows the graphs of f_A for some values of A belonging to that interval. We shall not deal with the problem of what f_A looks like for $A \notin [0, 4]$; after all, it is not difficult to find out. The important point is that in the latter case f_A does not map the interval $[0, 1]$ into itself. Let us remark that a simple substitution turns f_A into the function $g_A(x) = Ax(1 - x/a)$ which, for $A \in [0, 4]$, maps the interval $[0, a]$ into $[0, a]$ (a being any positive number).

The properties of sequences generated by the function f_A depend substantially on the parameter A. We will now show the nature of this dependence.

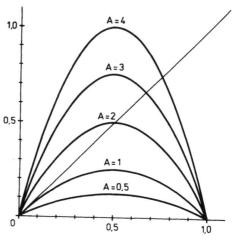

Figure 3.7

The first piece of information about the function f_A is provided by its fixed points. They are easily found by solving the equation $f_A(x) = x$, or

$$Ax(1 - x) = x.$$

We see that for $A \in [0, 1]$ there exists exactly one fixed point $\alpha = 0$ in $[0, 1]$, while for $A \in (1, 4]$ there are exactly two fixed points $\alpha = 0$, $\beta = 1 - 1/A$.

Let us now discuss in more detail the following three cases.

1. $A \in [0, 1]$. As the function has a unique fixed point $\alpha = 0$, its graph lies under the line $y = x$ (i.e. under the diagonal of the unit square). Therefore, $\alpha = 0$ is an attractive fixed point and for any $x_0 \in [0, 1]$ the sequence $f^n(x_0)$ converges to zero.

2. $A \in (1, 3]$. Observe what happens with the so far unique fixed point $\alpha = 0$ when A varies from 1 to 4. The point $\alpha = 0$ will 'split' — the point $\beta = 1 - 1/A$ will

separate and move away from it (for $A = 1$ we have $\beta = 0$ and for $A = 4$ we get $\beta = 3/4$). Thus the value $A = 1$ may be considered as critical for producing a second fixed point.

Applying Theorem 4, we easily verify that here zero is a repulsive fixed point and the fixed point $\beta = 1 - 1/A$, at least for $A < 3$, is attractive. (Readers may check this themselves.) For $A = 3$ we obtain $f_3'(\beta) = -1$, and so Theorem 4 is not applicable, but still β is an attractive fixed point. It is not so easy to prove that for $A \in (1, 3]$ the sequence generated by any x_0 converges, namely to β for $1 \neq x_0 \neq 0$. This follows from Theorems 6 and 7 that we shall state later (and implies, among other things, that for $A = 3$ the fixed point β is attractive). For the moment, readers may 'verify' this fact using a calculator; the convergence is rapid for $A = 2.5$, but becomes very slow for $A = 3$.

3. The case $A \in (3, 4]$ is the most dramatic. First of all, there always exists at least one pair of points $u, v \in (0, 1)$ with $f_A(u) = v$ and $f_A(v) = u$. The points u, v constitute a 2-cycle (i.e. a cycle of order 2) of f_A. For u and v we have $f_A^2(u) = u$ and $f_A^2(v) = v$. Hence, both u and v are fixed points of the function f_A^2 and can be found by solving the equation $f_A^2(x) = x$, that is

$$A[Ax(1 - x)][1 - Ax(1 - x)] = x$$

which reduces to

$$A^3x^4 - 2A^3x^3 + (A^3 + A^2)x^2 - A^2x + x = 0.$$

Both fixed points (of order 1) $x = 0$ and $x = 1 - 1/A$ solve the above equation. Hence, we may divide it by the polynomial $x[x - (1 - 1/A)]$, which gives

$$A^2x^2 - (A^2 + A)x + (A + 1) = 0. \tag{3.7}$$

The discriminant of the last equation is $D = A^4 - 2A^3 - 3A^2 = A^2(A + 1)(A - 3)$. We see that $D > 0$ (and hence the equation has two distinct roots) if, and only if, $A > 3$

(the case $A < -1$ has been excluded). For $A = 3$ we get $D = 0$ and the only root — a double one — is exactly the other fixed point $\beta = 2/3$. Thus the 2-cycle mentioned above is produced by separating from the fixed point $\beta = 2/3$ two other points (not fixed points) which will move away from β and from each other with growing A.

Equation (3.7) allows us to calculate the two points constituting the 2-cycle. For example, for $A = 3.1$ the approximate values obtained are $\gamma_1 = 0.76456...$ and $\gamma_2 = 0.55801...$. There is still another way of verifying the existence of a 2-cycle for the parameter value $A = 3.1$. Examine the values of the function f_A^2 at $x^1 = 0.7$ and $x^2 = 0.8$. We have

$$f_A^2(x^1) = 0.7043... > x^1, \qquad f_A^2(x^2) = 0.7749... < x^2.$$

Theorem 2 implies that there must be some fixed point γ of f_A^2 between x^1 and x^2. This, however, need not mean very much; in fact, γ could turn out to be a fixed point of f_A as well. Anyway, this is not the case, because the fixed points of f_A for $A = 3.1$ are $\alpha = 0$ and $\beta = 2.1/3.1 = 0.6774...$ and neither of them lies in the interval (x^1, x^2).

A certain value of $A > 3$ gives rise to the first 4-cycle (i.e. four distinct points y^1, y^2, y^3, y^4 with $f_A(y^1) = y^2$, $f_A(y^2) = y^3, f_A(y^3) = y^4, f_A(y^4) = y^1$). Later, the first cycle of order $2^3 = 8$ appears, still later the first cycle of order 16, etc. Starting with $A = A_c = 3.5700...$ we begin to observe cycles of orders $2^i p$ with $p > 1$, an odd number. And finally, at $A = 1 + 8^{1/2} = 3.8284...$ the first 3-cycle appears. For $A \geqslant 1 + 8^{1/2}$, the function f_A will already have cycles of all orders.

The reader may try to discover a 3-cycle for $A > 1 + 8^{1/2}$ similarly as the 2-cycle. It suffices to find two points z^1 and z^2 with $f_A^3(z^1) < z^1$ and $f_A^3(z^2) > z^2$ and such that no fixed point of f lies between z^1 and z^2. For

$A = 3.83187 \ldots$ one of the points of a 3-cycle will be exactly $\gamma = 0.5$. Figures 3.8 and 3.9 display the graphs of f_A^2 and f_A^3 for some values of the parameter A. They illustrate clearly how the cycles are produced.

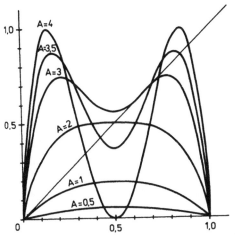

Figure 3.8

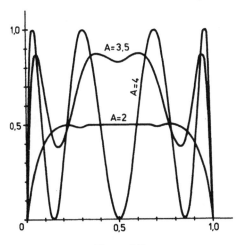

Figure 3.9

Using a more powerful computer (or theoretically, as we shall see later), the above-mentioned parameter value $A_c = 3.5700 \ldots$ can be shown to have a privileged position. Namely, if $A < A_c$, then every sequence generated by any $x_0 \in [0, 1]$ is either periodic (and its period is necessarily one of the numbers $1, 2, 2^2, 2^3, 2^4, \ldots$) or converging to such a sequence, i.e. for some periodic sequence $y_0, y_1, y_2, \ldots$ with period 2^n we have $\lim_{i \to \infty} |y_i - x_i| = 0$. Moreover, the said periodic sequence is generated by y_0 and f_A. We say that the sequence $\{x_n\}_{n=0}^{\infty}$ is asymptotically periodic. This behaviour of a sequence is still 'reasonable' enough; in our model of population growth it means that the number of individuals in remote generations varies approximately periodically.

On the other hand, if $A > A_c$, we can always find an initial point x_0 such that the sequence generated by it displays no regularity. The sequence (for a suitable x_0) is so much disordered that it may be taken for a sequence of random numbers (which is often done in practice). And that is what we call chaos.

The above example shows that it can be rather difficult to investigate the existence of cycles of a function. The properties of iterates of the function $Ax(1 - x)$, as well as other above-mentioned functions, have for a long time been an object of interest of many mathematicians. The first papers known to the author of this book date back to the 1930s. A survey can be found, for example, in [22]. Other properties have also been studied, besides the mentioned ones, such as the problem of stability of fixed points and cycles — we shall treat them later. However, only definite and not very complicated functions were considered. That is why, some ten years ago, a great sensation was caused by a paper by a Soviet mathematician A N Šarkovskii [34] which was published earlier in 1964 (at that time, none of the interested parties took notice of it) and which solved the problem of the existence of cycles in full generality. Here is the theorem.

Theorem 6 (Šarkovskii)

Let f be a continuous function mapping an interval I into I. In the set of positive integers, define a new ordering as follows

$$3 \prec 5 \prec 7 \prec \ldots \prec 2\cdot3 \prec 2\cdot5 \prec 2\cdot7 \prec \ldots \prec 2^i3 \prec 2^i5 \prec 2^i7$$
$$\ldots \prec 2^{j+1} \prec 2^j \prec \ldots \prec 8 \prec 4 \prec 2 \prec 1$$

(that is, first come all the odd numbers different from 1 in the natural order, then their doubles, then their quadruples etc, and finally the powers of 2 in decreasing order). If f has a cycle of order m and m $\prec$ n, then f has a cycle of order n as well.

According to several outstanding mathematicians, the above theorem is among the best that has been achieved in mathematics in the past twenty years. The reader expecting a proof of the theorem will be disappointed. The proof is too long. Moreover, Šarkovskii's original proof is hard to understand.

Fortunately, there are several lucid and relatively short proofs of Šarkovskii's theorem known today (let us mention at least [4], [5] and [37]).

For a better understanding of the properties of iterates, other results found by Šarkovskii are also very important. Some of them are detailed below.

Theorem 7

Let f be a continuous function mapping an interval I into I. Then the sequence generated by any x ∈ I and by f converges to some fixed point if, and only if, f has no cycle except cycles of order 1 [35].

The proof is again omitted, due to being very complicated. Yet we shall prove the following theorem, which is in fact a consequence of Theorem 7.

Theorem 8

Let f be a continuous function from an interval I into I and having cycles of orders $1, 2, 2^2, ..., 2^n$ only. Then every sequence $\{f^n(x)\}_{n=1}^{\infty}$ with $x \in I$ converges to some cycle (or a fixed point) of f; hence it is asymptotically periodic.

The proof of this is as follows. The function f^{2^n} has no cycles of orders higher than 1, it has only fixed points. This follows from the fact that whenever $x^1, x^2, ..., x^m$ constitute a cycle of order $m = 2^k$ with $k \leqslant n$, then $f^{2^n}(x^j) = x^j$ for $j = 1, 2, ..., m$. In fact, f^{2^n} is the 2^{n-k}th iterate of $f^m = f^{2^k}$ and each of the points x^j is a fixed point of f^m.

Let x be an arbitrary point. Let $x_j = f^j(x)$ for $j = 1, 2, ..., 2^n$. From Theorem 7, every one of the sequences

$$x_j, f^{2^n}(x_j), f^{2 \cdot 2^n}(x_j) = (f^{2^n})^2(x_j), f^{3 \cdot 2^n}(x_j), ...$$

(that is, generated by x_j and f^{2^n}) converges to some a^j, $j = 1, 2, ..., 2^n$, which means that

$$\lim_{s \to \infty} (f^{s \cdot 2^n}(x_j) - a_j) = 0.$$

The continuity of f implies also

$$\lim_{s \to \infty} (f(f^{s \, 2^n}(x_j)) - f(a^j)) = 0.$$

The left-hand side of the last equality reduces to

$$0 = \lim_{s \to \infty} (f^{s \cdot 2^n}(f(x_j)) - f(a^j)) = \lim_{s \to \infty} (f^{s \cdot 2^n}(x_{j+1})) - f(a^j).$$

Since at the same time

$$\lim_{s \to \infty} (f^{s \cdot 2^n}(x_{j+1})) = a^{j+1},$$

the uniqueness of the limit implies that $f(a^j) = a^{j+1}$ for $j < 2^n$. Analogously, we may prove that $f(a^{2^n}) = a^1$. Thus the points $a^1, ..., a^{2^n}$ form a cycle of the function f. If no points occur more than once in it, then it is a cycle of order 2^n; otherwise its order is a divisor of 2^n (the whole cycle is repeated an integral number of times in the sequence of 2^n points), which was to be proved.

Note that, using Theorems 6 and 7, it is easy to explain why the function $Ax(1 - x)$ from Example 3 generates only convergent sequences for $A \leqslant 3$. This follows from the fact that the said function, for $A \leqslant 3$, has been shown to have no 2-cycles and hence, by the Šarkovskii Theorem, it has no cycles of order greater than 1, so that Theorem 7 applies.

To conclude this section, let us examine attractive and repulsive cycles. The ideas are analogous to those attributed to fixed points (see § 3.2) and are very important in view of the last theorem, which says that sequences generated by a function may converge to its cycles.

Definition 3. Let f be a continuous function mapping an interval I into I and let $a_1, ..., a_k$ form a cycle of f of order k. Then this cycle is

1. attractive if at least one point of that cycle is an attractive point of f^k and

2. repulsive if every point of that cycle is a repulsive point of f^k.

Example 4. Using a calculator, we may check that the 2-cycle of the function $f_A(x) = Ax(1 - x)$ from Example 3 for the parameter value $A = 3.1$, consisting of points $\gamma_1 = 0.76456...$ and $\gamma_2 = 0.55801...$, is an attractive cycle. It suffices to choose a point x_0 sufficiently close to γ_1 (see how near) and perform iterations. The sequence generated by x_0 will be seen to converge to the said cycle and hence it is asymptotically periodic, having exactly two limit points γ_1

and γ_2. It follows that both γ_1 and γ_2 are attractive fixed points of f_2^2. This does not occur by chance, as the following theorem indicates.

Theorem 9

Let f be a continuous function mapping an interval I into I. Then every point of an attractive cycle of order k of f is an attractive fixed point of f^k.

To prove this, let α be a point of a cycle of f which is an attractive fixed point of f^k and let β be another point of the same cycle. Then $\alpha = f^s(\beta)$ for some $s < k$. Since α is an attractive fixed point of f^k, there exists a neighbourhood U of α such that for every $\eta \in U$ we have

$$\lim_{n \to \infty} f^{kn}(\eta) = \alpha. \tag{3.8}$$

As f^s is continuous and $f^s(\beta) = \alpha$, there is a neighbourhood V of β with $f^s(V) \subset U$.

Now let $\xi \in V$. Then $f^s(\xi) \in U$ and (3.8) implies that

$$\lim_{n \to \infty} f^s(f^{kn}(\xi)) = \lim_{n \to \infty} f^{kn}(f^s(\xi)) = \alpha. \tag{3.9}$$

Since the function f^{k-s} is continuous as well, we infer from (3.9) that

$$\lim_{n \to \infty} f^{kn}(\xi) = \lim_{n \to \infty} f^{k-s}(f^s(f^{k(n-1)}(\xi))) = f^{k-s}(\alpha) = \beta.$$

Hence, β is an attractive fixed point of f^k and the proof is complete.

Example 5. Consider once more the function introduced in Example 3. Examining the graph of its second iterate (see figure 3.8), we can see that the 2-cycle which was attractive for $A = 3.1$ becomes a repulsive cycle for $A = 4$. It is sufficient to find the corresponding point of the 2-cycle (one of the two points equals approximately 0.34) and to apply Theorem 4. The derivative of f_A^2 at that point is perceptibly less than -1.

Our graph does not show what happens in the case $A = 3.5$. The only thing that can be seen is that the absolute value of the derivative at the first point of the 2-cycle differs very little from 1. Therefore it is necessary to compute the derivative of f_A^2. The following theorem facilitates the calculation. With its help, it is rather easy to see that f_A has a repulsive 2-cycle for $A = 3.5$.

Theorem 10

Let f be a continuous function from an interval I to I and having a derivative at every point of I. Suppose that $\alpha_1, \ldots, \alpha_k$ is a k-cycle of f. Put $D = f'(\alpha_1)f'(\alpha_2)\ldots f'(\alpha_k)$. Then the said cycle is attractive if $|D| < 1$, and repulsive if $|D| > 1$.

The proof is as follows. We know that differentiation of a composite function obeys the rule $[g(h(x))]' = g'(h(x))h'(x)$. This rule, repeatedly applied to the composite $f^k(x)$, yields

$$[f^k(x)]' = f'(f^{k-1}(x))f'(f^{k-2}(x))\ldots f'(f(x))f'(x).$$

Now putting $x = \alpha_1$ and using the equalities $\alpha_2 = f(\alpha_1)$, $\alpha_3 = f(\alpha_2) = f^2(\alpha_1), \ldots$ together with Theorem 4, we obtain the assertion of Theorem 10.

Exercises

3.11 Decide whether the function $Ax(1 - x)$ has an attractive 3-cycle for some parameter value A.

3.12 Find the parameter value A for which the 2-cycle of the function $Ax(1 - x)$ changes from being attractive to repulsive.

3.13 Sketch the graph of a function having an attractive 3-cycle.

3.14 Sketch a function having a 3-cycle which is neither attractive nor repulsive.

3.15 Let f be a continuous function mapping an interval I into I and let $x \in I$. Denote by $L_f(x)$ the set of all limit points of the sequence $\{f^n(x)\}_{n=1}^{\infty}$ generated by x. Prove the following propositions:

1. The set $L_f(x)$ is closed.
2. $f(y) \in L_f(x)$ for every $y \in L_f(x)$ (that is, $f(L_f(x)) \subset \subset L_f(x)$).
3. If $L_f(x) = A \cup B$, where A and B are disjoint closed sets and A is bounded (hence compact), then $f(A) \cap B \neq \emptyset$.

3.16 Applying Exercise 3.15, part 3, prove the following generalisation of Theorem 5: if $L_f(x)$ is a finite set of k elements, then its elements form a k-cycle of f.

3.17 Prove that if I is a closed interval and f a continuous function mapping I into I, then $f(L_f(x)) = = L_f(x)$ for every $x \in I$.

*3.4 CHAOS

As we have seen in the preceding section, a continuous function having cycles of a finite number of different orders only (by the Šarkovskii Theorem they are necessarily cycles of orders $1, 2, ..., 2^n$, where n is a positive integer) generates only asymptotically periodic sequences (Theorem 8). But what happens if the function has cycles of other orders as well? Then the behaviour of the sequences may be very complicated, as will be shown by the following example.

Example 6. The function $f: [0, 1] \to [0, 1]$, defined for all t by $f(t) = 1 - |1 - 2t|$, has cycles of all orders. This can be verified even without employing the Šarkovskii Theorem. The function f is termed a 'tent function' and the graph of f^n is formed by 2^{n-1} identical isosceles triangles without bases (why?): figure 3.10 depicts the graphs of f, f^2 and f^3. Thus the function f^n for an arbitrary n has 2^n fixed points, which is more than the total number of fixed points of the functions $f, f^2, ..., f^{n-1}$ altogether. Therefore, f^n has a fixed point that is not a fixed point of any of the functions $f, ..., f^{n-1}$, hence it is a point of a cycle of order n of the function f.

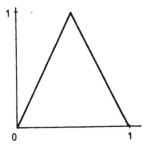

Figure 3.10a

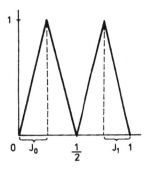

Figure 3.10b

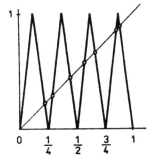

Figure 3.10c

For simplicity, put $J_0 = [0, 1/4]$, $J_1 = [3/4, 1]$ and $g = f^2$. It is easy to verify (see figure 3.10b) that

$$g(J_0) = g(J_1) = [0, 1]. \qquad (3.10)$$

The intervals J_0 and J_1 have the following interesting property: there exist points x and y such that all the terms of the generated sequences $\{g^n(x)\}_{n=0}^{\infty}$ and $\{g^n(y)\}_{n=0}^{\infty}$ lie in $J_0 \cup J_1$. Some of these sequences behave in a very irregular manner, as follows from the following lemma.

Lemma. *Given any sequence* $\{\alpha(n)\}_{n=0}^{\infty}$ *of zeros and ones, there exists a point* $x \in J_0 \cup J_1$ *such that* $g^n(x) \in J_0$ *if* $\alpha(n) = 0$, *and* $g^n(x) \in J_1$ *if* $\alpha(n) = 1$. *In other words,* $g^n(x) \in J_{\alpha(n)}$ *for every* n.

This can be proved. Since $g(J_0) \supset J_0 \cup J_1$, there exist closed intervals $J_{00}, J_{01} \subset J_0$ with $g(J_{00}) = J_0$ and $g(J_{01}) = J_1$. It is evident from figure 3.10b that the length of each of the intervals J_{00} and J_{01} equals a quarter of the length of J_0, that is $1/16$ (readers may easily verify that $J_{00} = [0, 1/16]$, $J_{01} = [3/16, 1/4]$). Similarly, we can find disjoint closed intervals $J_{10}, J_{11} \subset J_1$ with length $1/16$ and such that $g(J_{10}) = J_0$ and $g(J_{11}) = J_1$ (in particular, $J_{10} = [15/16, 1]$, $J_{11} = [3/4, 13/16]$).

As $g(J_{00}) = J_0$ and $J_{00}, J_{01} \subset J_0$, there are closed intervals $J_{000}, J_{001} \subset J_{00}$, each of them having the length $1/64 = 4^{-3}$ and satisfying $g(J_{000}) = J_{00}$, $g(J_{001}) = J_{01}$ (that is, $g^2(J_{000}) = J_0$, $g^2(J_{001}) = J_1$). Analogously, we may construct six other intervals $J_{010}, J_{011}, J_{100}, J_{101}, J_{110}$ and J_{111} of order 3 (their length being $1/64$).

We may continue in the same way. Corresponding to any finite sequence $\alpha(0), \alpha(1), \ldots, \alpha(n)$, we find a closed interval $J_{\alpha(0)\alpha(1)\ldots\alpha(n)}$ with length $1/4^{n+1}$ and such that

$$J_{\alpha(0)} \supset J_{\alpha(0)\alpha(1)} \supset \cdots \supset J_{\alpha(0)\alpha(1)\ldots\alpha(n)}$$

and

$$g(J_{\alpha(0)\alpha(1)\ldots\alpha(n)}) = J_{\alpha(1)\ldots\alpha(n)}.$$

Now consider the set

$$\bigcap_{n=0}^{\infty} J_{\alpha(0)\ldots\alpha(n)}.$$

It is an intersection of nested closed intervals with lengths tending to zero. As we know, such an intersection contains exactly one element, which we shall denote by x.

The point x has all the required properties. In fact,

$$x \in J_{\alpha(0)} \cap J_{\alpha(0)\,\alpha(1)} \cap J_{\alpha(0)\,\alpha(1)\,\alpha(2)} \cap \ldots \subset J_{\alpha(0)}$$

$$g(x) \in g(J_{\alpha(0)\,\alpha(1)}) = J_{\alpha(1)}$$

$$g^2(x) \in g^2(J_{\alpha(0)\,\alpha(1)\,\alpha(2)}) = g(J_{\alpha(1)\alpha(2)}) = J_{\alpha(2)} \quad \text{etc.}$$

The lemma is proved and we may continue discussing our example.

The sequence $\{\alpha(n)\}_{n=0}^{\infty}$ mentioned in the lemma will be referred to as the code of the point x. The lemma says that there are points u, v whose codes are

$$1\,0 \quad 1\,1\,0\,0 \quad 1\,1\,1\,0\,0\,0 \ldots \underbrace{1\ldots1}_{n}\underbrace{0\ldots0}_{n}\ldots$$

and

$$0\,0 \quad 0\,1\,0\,0 \quad 0\,1\,1\,0\,0\,0 \ldots \underbrace{0\,1\ldots1}_{n}\underbrace{0\ldots0}_{n}\ldots$$

The first of the above sequences is formed by groups always containing n ones followed by n zeros, $n = 1, 2, \ldots$, while in each group of the second sequence the first 1 is replaced by 0.

It is easy to verify that for every positive integer n there is m with

$$g^m(u), g^m(v) \in J_{\underbrace{0\ldots0}_{n}}.$$

(For instance, $g^4(u), g^4(v) \in J_{00}$, $g^9(u), g^9(v) \in J_{000}$, etc.) Since the length of the interval $J_{0\ldots0}$ of order n is 4^{-n}, we have, for such m,

$$|g^m(u) - g^m(v)| \leqslant 1/4^n. \tag{3.11}$$

Similarly, for $k = 0, 2, 6, 12, 20, \ldots$ we have $g^k(u) \in J_1$, $g^k(v) \in J_0$, and hence

$$|g^k(u) - g^k(v)| \geqslant 1/2. \tag{3.12}$$

From (3.11) and (3.12) we get

$$\limsup_{k \to \infty} |g^k(u) - g^k(v)| \geqslant 1/2$$

$$\liminf_{k \to \infty} |g^k(u) - g^k(v)| = 0$$

and hence also

$$\limsup_{k \to \infty} |f^k(u) - f^k(v)| \geqslant 1/2 \tag{3.13}$$

$$\liminf_{k \to \infty} |f^k(u) - f^k(v)| = 0. \tag{3.14}$$

The conditions (3.13) and (3.14) mean that the sequences generated by the points u and v and the function f are incomparable. Repeatedly they approach arbitrarily close to each other and then part (to a distance of at least $1/2$).

We may also verify that none of the sequences generated by u and v are asymptotically periodic, i.e. every periodic point p of f satisfies

$$\limsup_{n \to \infty} |f^n(u) - f^n(p)| \geqslant 1/4. \tag{3.15}$$

In fact, let k be the order of p. The point p is in one of the intervals $[0, 1/2)$, $[1/2, 1]$, say $p \in [0, 1/2)$. From the way we have chosen the code of u it follows that for some arbitrarily large n we have $f^{kn}(u) \in J_1$. Hence, $f^{kn}(u) \geqslant 3/4$, and, as $f^{kj}(p) = p < 1/2$ for every j, we have

$$|f^{kn}(u) - f^{kn}(p)| > 1/4$$

which implies (3.15). We can see that the behaviour of the

sequences generated by the function $f(t) = 1 - |1 - 2t|$ is very complicated.

The set $S = \{u, v\}$ described in the above example is termed a scrambled set of f. This concept is important and we give its formal definition below.

Definition 4. Let f be a continuous function from I to I, let $\varepsilon > 0$ and let $S \subset I$ be a set such that for each two distinct points $x, y \in S$ and for every periodic point p of f we have

$$\limsup_{n \to \infty} |f^n(x) - f^n(y)| > \varepsilon \qquad (3.16)$$

$$\liminf_{n \to \infty} |f^n(x) - f^n(y)| = 0 \qquad (3.17)$$

$$\limsup_{n \to \infty} |f^n(x) - f^n(p)| > \varepsilon. \qquad (3.18)$$

Then S is called an ε-scrambled set. If $\varepsilon = 0$, we simply say a scrambled set.

A function f is said to be chaotic if it has a scrambled set of at least two elements.

There are many chaotic functions, owing to the following proposition (see [12]).

Theorem 11

Let f be a continuous function from I to I and having a cycle of an order which is not a power of 2. Then, for some $\varepsilon > 0$, the function f has an infinite, closed ε-scrambled set S.

Proof of the theorem is complicated and therefore will be omitted. A weaker version of the theorem was proved in 1975 by Li and Yorke [20] and in this connection chaos in the sense of Li and Yorke is specifically mentioned; there are also other, non-equivalent ways in which chaos can be defined. The proof by Li and Yorke is also rather complicated.

Yet it is relatively easy to prove that every continuous function having a cycle of an order not equal to 2^n ($n = 0, 1, 2, ...$) has a two-point scrambled set. The key role is played by the following theorem, which is interesting in itself.

Theorem 12

Let g be a continuous function from I to I. Then g has a cycle of an order not equal to 2^n ($n = 0, 1, 2, ...$) if, and only if, there is a number m and two disjoint closed intervals $J_0, J_1 \subset I$ such that

$$g^m(J_0) \cap g^m(J_1) \supset J_0 \cup J_1 \qquad (3.19)$$

(see Šarkovskii [35]).

The theorem is easy to prove (see Exercise 3.23).

If g is a function having a cycle of the said type, then by Theorem 12 we may find a number m and disjoint intervals J_0, J_1 to satisfy (3.19) and apply to the function g^m the reasoning performed in Example 6 with g. We shall find that g^m is a chaotic function, determine its ε-scrambled set $S = \{x, y\}$ and easily check that S is an ε-scrambled set of the function g as well.

We have described two possible types of behaviour of functions so far, namely functions generating only asymptotically periodic sequences and chaotic functions. The question remains whether these are all the possible cases. The answer is negative. There is still another possibility between chaos and asymptotic periodicity.

Example 7. A detailed examination of the function $f_A(x) = Ax(1 - x)$, introduced already in § 3.3, reveals the following.

Denote by A_k the smallest number in $[0, 4]$ with the property that for $A > A_k$ the function f_A has a cycle of order 2^k. In §3.3 we have seen that $A_1 = 3$ and the Šarkovskii Theorem yields

$$A_1 \leqslant A_2 \leqslant A_3 \leqslant \ldots < 1 + \sqrt{8} = 3.8284\ldots.$$

(As we said in §3.3, for $A > 1 + \sqrt{8}$ the function f_A has a 3-cycle.) It can be proved by a rather difficult argument that all the inequalities above are strict and

$$A_1 < A_2 < \ldots < A_c = 3.5700\ldots$$

where A_c is the lowest upper bound of the numbers A_k, that is, $A_c = \lim_{k \to \infty} A_k$.

The function $g = f_{A_c}$ has some interesting properties. It has only cycles of orders 2^i ($i = 0, 1, 2, \ldots$) and for some $x \in [0, 1]$ the set $L_g(x)$ of all limit points of the sequence $\{g^n(x)\}_{n=1}^{\infty}$ is infinite. This means that the sequence generated by g and by x is not asymptotically periodic. Nevertheless, it can be proved that g is not a chaotic function.

However, the third possible type of behaviour of functions is not really very important. This is implied by the following theorem, which again will be stated without proof (see [31]).

Theorem 13

Let f be a continuous function from I to itself. Then exactly one of the following two situations may arise.

1. For some $\varepsilon > 0$, the function f has an infinite closed ε-scrambled set S.

2. Every sequence generated by f can be approximated by periodic points. In other words, for every $x \in I$ and any $\varepsilon > 0$ there is a periodic point p of f such that

$$\limsup_{n \to \infty} |f^n(x) - f^n(p)| < \varepsilon. \tag{3.20}$$

What part 2 really means is that no sequence generated by f can be distinguished from an asymptotically periodic sequence, although the sequence itself need not be asymptotically periodic, as follows from Example 7.

A function having a cycle of order 2^n for all non-negative integers n, and at the same time having no cycles of any other orders, is termed a function of the type 2^∞.

In Example 7, we have introduced a non-chaotic function of the type 2^∞, which also generates sequences that are not asymptotically periodic. Examples are known of 2^∞-type functions generating only asymptotically periodic sequences (see Šarkovskii [35]). But functions of the said type may also be chaotic.

Example 8. For $\lambda \in [0, 1]$, let f_λ be a function mapping $[0, 1]$ into $[0, 1]$, defined for all t by

$$f_\lambda(t) = \min\{\lambda, 1 - |1 - 2t|\}.$$

Its graph is sketched in figure 3.11. As in Example 7, it is possible to construct a sequence

$$\lambda_1 < \lambda_2 < \dots < \lambda_k < \dots$$

with the property that f_λ has a cycle of order 2^k if, and only if, $\lambda > \lambda_k$. It can be proved that $\lim_{k \to \infty} \lambda_k = \lambda_c = 0.825\dots$ and that f_{λ_c} is a chaotic function of the type 2^∞ (see [24]).

Exercises

3.18 Show that the function $f(t) = 1 - |1 - 2t|$ from Example 6 has two cycles of order 3 and compute their points.

3.19 As an illustration of the lemma from Example 6, find explicitly a point $x \in [0, 1]$ such that $g^n(x) \in J_0$ for

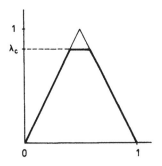

Figure 3.11

every $n \neq 4$ and $g^4(x) \in J_1$. (The notation used is that introduced in the proof of the lemma.)

3.20 A slight modification of the argument used in Example 6 makes it possible to prove that the function $f(t) = 1 - |1 - 2t|$ has an infinite (even uncountable) ε-scrambled set S with $\varepsilon = 1/4$. Prove that this is so. (Hint: S will contain the points corresponding to the codes of the form
$\square 0 1 \square 0 0 1 1 \square 0 0 0 1 1 1 \square 0 0 0 0 1 1 1 1 \square$....
The boxes should be filled with zeros and ones in such a way that the codes corresponding to any two points of S differ at infinitely many places (why?).)

3.21 To every code there corresponds exactly one point x satisfying the conditions of the lemma in Example 6. Prove it. (Hint: every interval $J \subset J_0 \cup J_1$ is mapped by g onto an interval four times longer.)

3.22 Let f be a continuous function from an interval I to I. Suppose that there exist disjoint intervals $J_0, J_1 \subset I$ satisfying $f(J_0) \supset (J_0 \cup J_1)$ as well as $f(J_1) \supset (J_0 \cup J_1)$, then show that f has a 3-cycle.

3.23 1. Give an example showing that the converse assertion to that stated in Exercise 3.22 is false.

2. Prove that if a continuous function g mapping an interval I into I has a cycle of order 3, then there are disjoint closed intervals $J_0, J_1 \subset I$ satisfying (3.19) with $m = 2$.

3. Prove Theorem 12.

3.5 STABILITY

Theorem 13 implies that it is crucial when using a mathematical model to know whether the function used in it is chaotic or not, as the character of the generated sequences in the former case is completely different from that in the latter. Therefore, the essential question for modelling which has already been mentioned in the introductory section, is: If, instead of the right function f, another function $\tilde{f}$ differing only a little from f is used in a model, can the sequences generated by $\tilde{f}$ have substantially different properties to those generated by the right function? If some property does not change essentially with small variations of f, the function f is said to be stable with respect to that property; otherwise it is unstable. In order to understand the problem better, we first give some examples.

Example 9. Consider the function f shown by the full line in figure 3.12a. It has two fixed points α and β. Examine what may happen when f is changed a little and replaced by the function $\tilde{f}$ shown by the dotted line. The point α may be pushed a little to the left or to the right — we shall have a fixed point $\tilde{\alpha}$ — but the point β may completely disappear, as in our case. It suffices that the values of f become lower everywhere in some neighbourhood of β. However, if the function value at β became higher (that is, we would have $\tilde{f}(\beta) > \beta$), then the new function $\tilde{f}$ would have two new fixed points $\tilde{\beta}_1$ and $\tilde{\beta}_2$ instead of the single fixed point β (figure 3.12b). In this case, α is a stable fixed point of f, while β is its unstable fixed point.

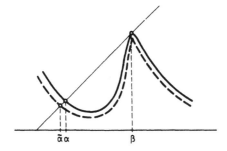

Figure 3.12*a*

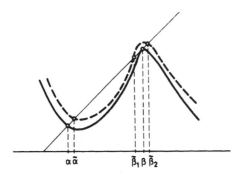

Figure 3.12*b*

Example 10. Another function f is shown by the full line in figure 3.13. It is defined by $f(x) = -x$ for $x \in [-1, 1]$ (we have already encountered it). The point 0 is its stable fixed point. In fact, if $|f(x) - \tilde{f}(x)|$ is sufficiently small, say if

$$|f(x) - \tilde{f}(x)| < \frac{1}{n}$$

for all x, then

$$\tilde{f}\left(\frac{1}{n}\right) < f\left(\frac{1}{n}\right) + \frac{1}{n} = 0$$

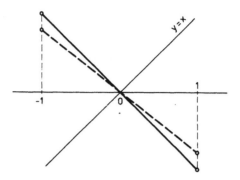

Figure 3.13

and

$$\tilde{f}\left(-\frac{1}{n}\right) > f\left(-\frac{1}{n}\right) - \frac{1}{n} = 0$$

and it suffices to apply Theorem 2: the function f has a fixed point between $-1/n$ and $1/n$. Also, every $x \neq 0$ is a point of a 2-cycle of f, but the 2-cycle is unstable. To see this, consider the function $f(x) = -\lambda x$, where λ is any positive number less than 1 (in figure 3.13 it is shown by a dotted line). The said function has no cycles, it has exactly one fixed point and generates only convergent sequences.

Example 11. Let us recall the function $f_A(x) =$ $= Ax(1 - x)$ introduced in Example 3. As we have seen, it has a 3-cycle for $A \geqslant 1 + \sqrt{8}$, but not for smaller values of A. This means that for $A = 1 + \sqrt{8}$ the function f_A has an unstable 3-cycle; if A decreases by an arbitrarily small amount, the obtained function $\tilde{f}$ will be close to f_A but will have no 3-cycle. On the other hand, by increasing A by an arbitrarily small amount, the said unstable 3-cycle will split and produce two cycles of order 3 (similarly to the way the fixed point β produced two fixed points $\tilde{\beta}_1$ and $\tilde{\beta}_2$ in Example 9).

We have now come to the interesting problem of whether an arbitrarily small change can deprive a function of all its cycles. The problem was solved in 1981: it is possible if the function has cycles of order 2 only, all of them being unstable, and fixed points. We state the theorem in full shortly. Let us point out that up to the end of this section all the functions considered will be defined on a closed bounded interval I.

Theorem 14

Let f be a continuous function mapping a closed bounded interval I into I and having a cycle of order m. Then there exists a number $\varepsilon > 0$ such that every continuous function $g: I \to I$ satisfying $|f(x) - g(x)| < \varepsilon$ for all $x \in I$ has a cycle of order k, where k is any positive integer preceded by m in the Šarkovskii ordering (see Block [3]).

Proof of the theorem is omitted due to being very complicated. It follows immediately from this theorem that by a slight modification of a function f having a cycle of order $2^s p$, where p is an odd number greater than 1 (such a function is chaotic by Theorem 11), we obtain a function g having a cycle of order $2^s(p + 2)$ and hence chaotic as well. Some chaotic functions may become non-chaotic by an arbitrarily small change; however, this property is displayed only by functions of the type 2^x, and they are not many, as can be shown. We infer that chaotic functions are 'essentially' stable.

What about the stability of non-chaotic functions? The answer is not very encouraging.

Theorem 15

For each continuous function f mapping a closed bounded interval I into I and for any $\varepsilon > 0$ there exists a chaotic function g with $|f(x) - g(x)| < \varepsilon$ for all $x \in I$.

In other words, by an arbitrarily small change we may obtain a chaotic function from any non-chaotic one (see Kloeden [13]).

The proof is simplified for the case $I = [0, 1]$. The continuity of f implies that f has a fixed point $\alpha = f(\alpha)$ in I. Also, there exists an interval $J = (a, b) \subset I$ containing α and such that $|f(x) - \alpha| < \varepsilon$ for all $x \in J$. (Here we have slightly simplified the situation by ignoring the cases $\alpha = 0$ and $\alpha = 1$. Readers will certainly be able to work out how to deal with those cases.)

Now consider the rectangle K with vertices $[a, \alpha - \varepsilon]$, $[b, \alpha - \varepsilon]$, $[b, \alpha + \varepsilon]$ and $[a, \alpha + \varepsilon]$. In its interior, we may construct a square H with its centre at $[\alpha, f(\alpha)]$ — see figure 3.14. The square H can be expressed as a Cartesian product

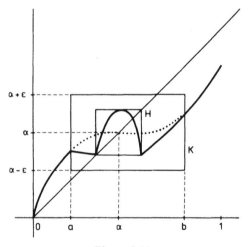

Figure 3.14

$H = V \times V$, where V is some interval with centre at α. Now modify the function f so that the part of its graph inside the square H is replaced by the graph of a chaotic function, e.g. by a reduced copy of the graph of the chaotic function

$4x(1 - x)$ examined in Example 3. Outside the square H, but still inside the rectangle K, we connect the graph of that chaotic function continuously with the graph of the original function f so that $g(a) = f(a)$ and $g(b) = f(b)$. (See figure 3.14 for details.) The continuous line represents the new function g, the dotted line shows that part of the graph of the given function which was modified. For the construction to be correct it is essential that $g(V) \subset V$. It is not difficult to test that the new function g has a 3-cycle and to apply Theorem 11.

However, the situation does not seem to be so difficult in most cases. It is true that a slight change in a non-chaotic function may produce a chaotic one, but the chaos will be small. If the original function has no cycles and the set of its fixed points does not include any interval, then a small change may only produce a function generating almost convergent sequences (see [33]).

Theorem 16

Let f be a continuous function from a closed bounded interval I to I and assume that the set of its fixed points includes no interval. Then, given any $\varepsilon > 0$, every continuous function $g: I \rightarrow I$ produced from f by a sufficiently small change satisfies

$$\left| \limsup_{n \to \infty} g^n(x) - \liminf_{n \to \infty} g^n(x) \right| < \varepsilon.$$

This theorem can be generalised [39].

Another way to deal with the problem is not to allow the derivative to change very much, i.e. suppose that f differs little from g and, at the same time, the derivative f' differs little from g' (the functions f and g in the proof of Theorem 15 do not satisfy that condition). Further interesting points concerning cycles and stability are mentioned in the exercises.

Exercises

3.24 Let f be a function mapping $[0, 1]$ into $[0, 1]$ and satisfying $f(0) = 1/2, f(1/2) = 1, f(1) = 0$ (hence, f has a 3-cycle). Complete the definition of f in such a way that the said cycle is

1. stable,
2. unstable.

3.25 Find a function having an unstable 4-cycle.

3.26 Given a function f having a cycle of order k, the method illustrated by figure 3.15 allows us to construct a function g having a cycle of order $2k$. The big square is a Cartesian product $J \times J$, where J is an interval. The two small squares inside are identical. One of them

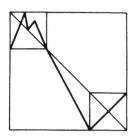

Figure 3.15

contains the graph of f, reduced as necessary. The graph of the new function g, mapping J into J, is formed by the graph of f and two straight segments.

1. Verify that the construction of the function g is correct.

2. Using the described construction, define a continuous function having an unstable 8-cycle.

3.27 A function f which has a 5-cycle but not a 3-cycle (and is continuous) must map the elements of its 5-cycle $a_1 < a_2 < a_3 < a_4 < a_5$ either in the sense indicated by arrows in figure 3.16, that is $f(a_3) = a_4, f(a_2) = a_5$, etc, or all in the opposite sense (that is $f(a_3) = a_1, f(a_2) = a_4$, etc). Sketch the graph of a continuous function having a 5-cycle but no 3-cycle. Use both ways of defining the function at the points of the cycle.

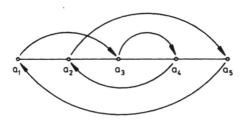

Figure 3.16

3.6 CONCLUDING REMARKS

When investigating iterates of functions, a calculator is a valuable tool. Readers will have appreciated it while reading this chapter. In fact, without a calculator it is almost impossible to localise, for instance, the 7-cycle of the function $4t(1 - t)$, which is known to exist. Also, it is almost impossible to find, without a calculator, the critical parameter value for the first 7-cycle to appear. The search for cycles of higher orders necessitates a computer. The use of a computer, in its turn, helps to discover various interesting results, like the following fascinating result, discovered a few years ago.

As we have seen in Example 3, the first 2-cycle of the function $At(1 - t)$ appears at the parameter value $A = 3$, the first 4-cycle arises at some $A_2 > A_1$ and so on. Thus we obtain a sequence of numbers $A_1 < A_2 < A_3 < \ldots$ in which A_n is the critical parameter value for the first 2^n-cycle of the said function to appear. Physicist M Feigenbaum noticed that the sequence $\{A_{n+1} - A_n\}_{n=1}^{\infty}$ of differences of consecutive critical values behaves, for large n, approximately like the geometrical sequence with quotient $1/\delta$, where $\delta = 4.669\,201\ldots$ (a long queue of further digits follows). More exactly, if

$$\delta_i = (A_i - A_{i-1})/(A_{i+1} - A_i)$$

then the sequence δ_i converges to the number δ. This might not seem very surprising. However, Feigenbaum has found that other parametrised systems of functions display the same behaviour. And the limit is always the same constant δ. Thus the result published in 1978 [10] had not been proved by a classical mathematical argument but was believed to hold for all parametrised systems of the type $Ag(t)$, where g is a continuous function mapping $[0, 1]$ into $[0, 1]$ and satisfies the conditions:

1. $g(0) = g(1) = 0$ and $g(t) > 0$ for $t \in (0, 1)$;
2. g has a continuous derivative in $[0, 1]$;
3. g has exactly one maximum, at a point ξ at which it has a second derivative;
4. g is increasing in the interval $[0, \xi]$ and decreasing in $[\xi, 1]$.

The result was correctly proved as late as 1982, still with the aid of a computer. (See also [5a] and [20a].)

Note that it is only with the use of computers that certain problems concerning iterations can be solved today. This applies in particular to iterates of functions of several variables (an example of such a function will be given in

Chapter 4). The reason is that while the properties of iterates of functions of one variable — at least the most simple ones — are to a certain extent theoretically mastered, almost nothing is known about iterates of functions of several variables. This, however, does not mean that no problems are left in the one-dimensional case. The results stated in §§ 3.4 and 3.5 point to some of them. They involve, in particular, the questions about stability of functions. What changes in cycles, their orders, and other properties are brought about by slightly changing the original function? From the practical point of view, we are only interested in perceptible cycles, i.e. those in which the distance between the largest and the smallest element is greater than some positive number δ. Smaller cycles cannot be distinguished from noise. It is found, for instance, that Theorem 14 (by Block) is not very good in that connection (see [32]). Another problem is whether a small change of a slightly chaotic function produces a very chaotic function. How large can the chaotic set of the new function be? (See [23] and [30].)

Still the most difficult problems are those arising in connection with functions of several variables. For example, the Šarkovskii Theorem is known to break down even for continuous functions mapping $I \times I$ into $I \times I$, where I is a closed interval. In fact, it can be shown that for any set A of positive integers, with $1 \in A$, there is a continuous function mapping $I \times I$ into $I \times I$ and having just cycles of order n for all $n \in A$. It is still possible that those functions obey a Šarkovskii-type rule concerning the existence of cycles under some additional assumptions.

Some other problems, together with a survey of known results and a detailed bibliography of the already vast amount of literature, can be found in several monographs that have appeared recently. We point out especially [6] and [38].

4. APPLICATIONS OF ITERATIONS

4.1 INTRODUCTION

The renaissance of interest in iterations is closely connected, as we have already mentioned in the introduction to Chapter 3, with their applicability in physics and biology, and also in other sciences. They can be employed to model the growth and evolution of certain systems or situations. However, the study of the behaviour of such models, especially in physics, often presents a very difficult problem which can only be solved today with the use of powerful computers. The problem is so difficult because the state of a typical physical system at the time $t = n$ ($n = 0, 1, 2, \ldots$ years, days, hours, seconds etc) cannot be described by a single variable x_n. A k-tuple of variables $X_n = (x_n^1, x_n^2, \ldots, x_n^k)$ is needed, k often being very large. The equation describing the behaviour of the system has the form

$$X_{n+1} = F(X_n)$$

where F is a function of k variables. However, most of the known theoretical results concerning the properties of iterates deal only with functions of a single variable, as we have already mentioned at the end of Chapter 3.

Thus, this chapter is devoted to applications of iterations in biology, where in many cases we can manage with continuous functions of one variable. We shall suppose that readers have read Chapter 3 and refer to it at times.

The simplest mathematical model of the population growth of a single species of living organisms is obtained by counting all the individuals of that species at given times. Thus we get several terms of a sequence of positive numbers (or finish by getting a zero if the species dies out). If we succeed in determining how the sequence terms depend on the preceding terms, we may define a sequence $\{x_n\}$ whose terms represent the population of the given species in the given territory at the time $n = 1, 2, 3, \dots$. Taking the limit as $n \to \infty$, we may study the dynamics of the population.

Some 800 years ago, Leonardo of Pisa, surnamed Fibonacci, set up a model for the growth of a population of rabbits. Let us follow his line of thought. Every pair of rabbits, at the time $t = 1$ as well as at the time $t = 2$, gives birth to two young ones each time, one of them being always a male and one a female. The initial time $t = 0$ is the birth time of the first pair of rabbits. Denote by x_n the number of newborn pairs of rabbits at the time $t = n$. We have

$$x_0 = x_1 = 1, \quad x_n = x_{n-1} + x_{n-2} \quad \text{for} \quad n > 1.$$

The above equalities define the Fibonacci sequence

$$x_n = \frac{1}{\sqrt{5}} \left[\left(\frac{1 + \sqrt{5}}{2} \right)^{n+1} - \left(\frac{1 - \sqrt{5}}{2} \right)^{n+1} \right]$$

(as can be verified by mathematical induction). It is interesting that

$$\lim_{n \to \infty} \frac{x_{n+1}}{x_n} = \frac{1 + \sqrt{5}}{2},$$

which means that there exists a state of equilibrium in the described population. For large n the sequence will behave approximately like the geometrical sequence with quotient

$(1 + \sqrt{5})/2$. Of course, this is true provided that the rabbits live in conditions that do not restrict their reproduction (sufficient food, space, no predators etc), which is never fulfilled in reality. Let us note that the above model involves iterates of a function of two variables, hence it is not a one-dimensional case.

In the next section we shall examine the whole idea a little more exactly. We shall also take into account, in the mathematical model, that the population growth is restricted.

4.2 A DISCRETE MODEL OF POPULATION GROWTH FOR A SINGLE BIOLOGICAL SPECIES

Let us begin by introducing the necessary concepts. We wish to establish a mathematical model describing the evolution of a population of a single biological species in time, or, expressed in biological terminology, a one-species population growth model. Here, the population means the system of the living organisms in a given territory.

Example 1. Suppose that there are x_0 individuals of a given species living at time t_0 in a given territory. To simplify the notation, we may put $t_0 = 0$. At the time $t = 1$ (year, day, or hour), the number of individuals will be x_1. Evidently, the number x_1 can be obtained by subtracting from x_0 the number of individuals that died between the times 0 and 1, and by adding the number of newborn ones, so

$$x_1 = x_0 + \frac{k}{100} x_0 - \frac{s}{100} x_0$$

where k and s are positive constants giving the increment and the loss in percentage of the number x_0, respectively. In short, we may write

$$x_1 = x_0 q.$$

The number q may be referred to as the coefficient of growth ($q > 1$ represents an increase, and $q < 1$ a decrease in population).

Assuming that q does not depend on x_0, we get

$$x_2 = x_1 q = x_0 q^2$$

and in general, for any positive integer n, we have

$$x_n = x_{n-1} q = x_0 q^n. \tag{4.1}$$

Thus, $\{x_n\}$ is a geometrical sequence. If $q > 1$, the population is growing beyond all limits, while for $q < 1$ we have the opposite situation — the population is dying out. In the latter case, $\lim_{n \to \infty} x_n = 0$. The latter case may correspond to reality, but the former can never happen, as the resources of growth are limited for every population. Therefore, if we want to model populations other than those dying out, the coefficient of growth must not be supposed constant, but rather a function of the population. This means that a population x grows or decreases in number with a coefficient $q = q(x)$.

And so, supposing that the given environment has a certain capacity $\bar{x}$, i.e. that for a prolonged time it cannot provide more than $\bar{x}$ individuals of the given species with what they need for life, then we may put

$$q = q(x) = 1 + A(\bar{x} - x)$$

where $A > 0$ is a constant. Thus q is proportional to the difference between the population x and the environmental capacity. Then, from (4.1), we have

$$x_{n+1} = x_n[1 + A(\bar{x} - x_n)].$$

This equation takes its simplest form when putting $\bar{x} = (A - 1)/A$. In our model that may mean, for $A \in (0, 1)$, that $\bar{x} < 0$, but that does not matter; the model we obtain,

$$x_{n+1} = Ax_n(1 - x_n) \qquad (4.2)$$

describes the situation adequately, even if $A \in (0, 1)$. In fact, in such a case we have $\lim_{n \to \infty} x_n = 0$ and the population is on its way to extinction. Let us note that our considerations have resulted in a model in which x means the relative population, being always a number in the interval $[0, 1]$.

Equation (4.2) has been analysed in Chapter 3 (see Example 3 in particular). We recall that for $A \in [0, 3]$ the function $Ax(1 - x)$ always has an attractive fixed point, namely $\alpha = 0$ if $A \in [0, 1]$, and $\beta = (A - 1)/A (= \bar{x})$ if $A \in (1, 3)$. Also, for the said values of the parameter A, the function $Ax(1 - x)$ has no cycles of higher orders. This means that a population governed by our model for $A \in [0, 3]$ is stable — the number of individuals approaches, in time, an equilibrium (i.e. a fixed point of the function $Ax(1 - x)$).

As we know, the function $Ax(1 - x)$ for $A > 3$ has cycles. But even then it may describe population dynamics. A good example is presented in [25]. The common vole population in Czechoslovakia is well known to develop periodically. Over-population, which occurs every 3 to 4 years, with as many as 1500 voles per hectare in a field, causes fights for space, food and shelter between individual colonies of voles. The animals live in permanent stress. They are exhausted after fights, have little resistance to sickness and, as a rule, they all die the next winter. The only ones to survive are those individuals that had migrated to inconvenient places (from their point of view); they will now return to the depopulated fields (the density of population being one vole in up to five hectares) and the whole cycle begins again.

The above model, for every value $A \in [0, 3]$, had just one attractive fixed point, i.e. only one state of equilibrium. The threshold model we are going to present now has two attractive fixed points separated by a repulsive fixed point.

Example 2. Let us investigate what happens if the growth function has the form

$$q(x) = Ax(1 - x).$$

It is not difficult to see that for $A \in [0, 27/4]$, the function $g_A(x) = [Ax(1 - x)]x = Ax^2(1 - x)$, describing the population growth, maps continuously the interval $I = [0, 1]$ into itself. It is sufficient to find the maximum of g_A.

The population then develops according to the equation

$$x_{n+1} = Ax_n^2(1 - x_n). \tag{4.3}$$

The fixed points of g_A are found by solving the equation $g_A(x) = x$, which reduces to

$$x(Ax^2 - Ax + 1) = 0. \tag{4.4}$$

Since $x = 0$ is a trivial solution, our function has a fixed point $\alpha = 0$ for all parameter values A. It is easy to check that g_A has a zero derivative at the fixed point α and using Theorem 4 of the last chapter, α is an attractive fixed point, whatever the value of A may be.

In order to find other possible fixed points, let us solve the equation

$$Ax^2 - Ax + 1 = 0 \tag{4.5}$$

obtained from (4.4), depending on the parameter A.

1. If $A < 4$, then the discriminant of (4.5) is negative, and hence $\alpha = 0$ is the only fixed point of g_A. It is easy to verify that for all $x > 0$ we have $g_A(x) < x$, and hence every sequence $\{x_n\}_{n=1}^{\infty}$ generated by any $x_0 \in [0, 1]$ converges to α.

2. For $A = 4$, the discriminant $D = A^2 - 4A$ of equation (4.5) equals zero and the equation has exactly one root

$\beta = 1/2$, which is the second fixed point of g_A. The above-mentioned Theorem 4 in this case gives no answer to the question of whether the fixed point β is attractive or repulsive, as $g'_4(\beta) = 1$. The situation is clearly seen from figure 4.1 depicting the graphs of g_A for $A = 3$, $A = 4$ and $A = 16/3$. The diagram, as well as a direct computation, shows that $g_4(x) < x$ whenever $x < \beta$, $x \neq 0$, and so β is not an attractive fixed point. Analogously, if $x > \beta$ and x is sufficiently close to β, then the sequence generated by x converges to β, hence β cannot be a repulsive fixed point either.

3. For $A > 4$ the fixed point $\beta = 1/2$ splits and is replaced by two fixed points

$$\beta_1 = \frac{A - (A^2 - 4A)^{1/2}}{2A} = \frac{1}{2} - \left(\frac{1}{4} - \frac{1}{A}\right)^{1/2},$$

$$\beta_2 = \frac{1}{2} + \left(\frac{1}{4} - \frac{1}{A}\right)^{1/2}.$$

We can obtain these by solving (4.5). It is easy to verify that $0 < \beta_1 < \beta_2 < 1$ and $g'_4(\beta_1) > 1$. Thus β_1 is a repulsive fixed point for any value $A \in (4, 27/4]$.

What about the second fixed point β_2? By computing $g'_4(\beta_2)$ we get

$$g'_4(\beta_2) = 3 - \frac{1}{2}A - \frac{1}{2}(A^2 - 4A)^{1/2}.$$

It can be computed that $g'_4(\beta_2) < 1$ for all A, but $g'_4(\beta_2) > -1$ if, and only if, $A < 16/3$. Thus, for $A \in (4, 16/3)$, β_2 is an attractive fixed point of g_A. A deeper analysis shows that β_2 is an attractive fixed point also if $A = 16/3$, although in that case $g'_4(\beta_2) = -1$. As we cannot recognise it from figure 4.1, we have to verify that g_A for $A = 16/3$ has no 2-cycle (this is left as an exercise for readers), then from the

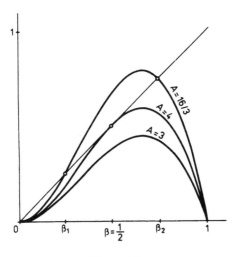

Figure 4.1

Šarkovskii Theorem it is found that g_A has no cycles at all
and it suffices to apply Theorem 7 of the preceding chapter.
For $A > 16/3$ we have $g'_A(\beta_2) < -1$, hence β_2 is a repulsive
fixed point. Moreover, for these parameter values, g_A also
has a 2-cycle (its two points have separated from the fixed
point β_2). For $A > 5.76\ldots$, the first 4-cycle appears, etc, and
beginning at $A = 5.89\ldots$ the function becomes chaotic.

Let us now examine in more detail the role of the fixed
points β_1 and β_2 for $A > 4$. As $g_A(1) = 0$, evidently there
exists a point $\gamma \in (\beta_2, 1)$ with $g_A(\gamma) = \beta_1$. If $\{x_n\}$ is the
sequence generated by x_0, then

$$\lim_{n \to \infty} x_n = 0 \quad \text{if} \quad x_0 \in [0, \beta_1) \cup (\gamma, 1] \qquad (4.6)$$

$$\lim_{n \to \infty} x_n = \beta_1 \quad \text{if} \quad x_0 = \beta_1 \quad \text{or} \quad x_0 = \gamma. \qquad (4.7)$$

The property (4.6) follows from the fact that $g_A(x) < \beta_1$ for
all $x \in [0, \beta_1) \cup (\gamma, 1]$, because g_A is increasing on $[0, \beta_1]$ and
decreasing on $[\gamma, 1]$. The condition (4.7) is evident.

Let us look again at the interval (β_1, γ). It can be found that there is a critical value $A_c = 6.6\ldots$ of the parameter A such that

$$g_A(x) \in (\beta_1, \gamma) \quad \text{whenever} \quad x \in (\beta_1, \gamma) \quad \text{and} \quad A \in (4, A_c). \qquad (4.8)$$

Now (4.6), (4.7) and (4.8) imply that β_1 and γ are threshold values. If the population does not attain β_1 at least, then it will die out. This occurs rather often, a small population is likely to become extinct under the influence of various disturbing factors. If the population attains exactly the value β_1, it is in equilibrium, yet only theoretically; the equilibrium is in fact very unstable. A small disturbance suffices to push it away from β_1 to one side or the other.

If the population exceeds the second threshold value γ, the same thing will happen as in the first case — it will die out. This is often the case in reality. Imagine an overpopulation of some herbivorous animals in a desert with very poor vegetation. This will result in extinction of the vegetation (which cannot reproduce itself at a sufficient rate and is absorbed by the desert) and, as a consequence, the animals will die out too.

If the population attains a value in the interval (β_1, γ), it tends, for $A \leqslant 16/3$, to the equilibrium β_2, which is stable; when removed from that state slightly by some disturbing factor, the population will return to it after some time. In the case that $A \in (16/3, A_c)$, the population may behave variously, even chaotically, but cannot die out. It varies between β_1 and γ. If A exceeds the critical value A_c, however, the population may die out in this case as well. We can see that the model (4.3) is nearer to reality than (4.2), but its dynamics are much more complicated and it is more difficult to analyse.

Concluding this section, let us mention two more models of one-species population growth which are most frequently encountered in the literature. One of them is the model

$$x_{n+1} = ax_n - bx_n^2 \qquad (4.9)$$

which in fact is just another version of model (4.2). Here, $a > 1$ and $b > 0$ are constants. This model may be written also in the form

$$x_{n+1} = x_n[1 + r(1 - x_n/K)] \qquad (4.10)$$

where $r > 0$ and $K > 0$ are constants. A non-trivial equilibrium of that model is $\bar{x} = K$. This means that K is the capacity of the environment and r is the growth coefficient. By a suitable substitution, (4.9) can be reduced to (4.2).

A model may also employ the exponential function

$$x_{n+1} = x_n e^{r(1 - x_n/K)}. \qquad (4.11)$$

Such a model has equilibria with properties similar to those previously described. Both models will be re-examined in the exercises.

Exercises

4.1 Show that the model introduced in Example 2 has two thresholds, even for $A = 4$. Explain how the model's behaviour depends on the choice of the initial point x_0.

4.2 In this exercise, g_A will denote the function $Ax^2(1 - x)$ from Example 2.

1. Show in detail that, for $A \in (4, 16/3)$, β_2 is an attractive fixed point.
*2. Show that for $A = 16/3$ the function g_A has no 2-cycle. (Hint: for $A = 16/3$, the equation $g_A^2(x) = x$ has a triple root $\xi = \beta_2 = 3/4$.)

3. Using a calculator, for $A = 5.5$, find both points of the (only) 2-cycle of g_A. Is this cycle attractive?

4. Estimate the critical value A_c of the parameter of g_A to three significant figures accuracy.

4.3 Analyse model (4.10), i.e. explain how the population growth obeying this model depends on the choice of the initial point x_0, for various parameter values.

4.4 Repeat Exercise 4.3 with model (4.11).

4.3 A DISCRETE MODEL OF EPIDEMICS

The basic assumption for establishing a simple model for the propagation of epidemics is the following. Each individual has the same number of contacts with other individuals in a given period of time, that is an equal chance of getting infected with the disease. Moreover, we shall suppose that the population is constant, i.e. the number of individuals considered does not change. The period during which a sick person is contagious is also supposed to be constant and its length equals the unit of time (a week, 10 days or so, depending on the kind of disease). After overcoming the disease, depending again on its kind, the individual may either remain permanently immune, or return without immunity to the group of susceptible people. Let us first examine the second possibility and begin with the simplest case.

Example 3. Divide the population into two groups. Let $N(t)$ denote the number of infected persons at time t, and $V(t)$ the number of susceptible (i.e. healthy) individuals who can get infected at the time t, the total population being X. Suppose also that if two individuals are arbitrarily chosen from the population, one of them being healthy and the other sick, then the probability of the healthy person getting infected from the sick one within a unit of time is p and does not depend on the choice of the two individuals. Also, denote $q = 1 - p = e^{-a}$. Since $0 \leqslant p \leqslant 1$, we have $0 \leqslant q \leqslant 1$ as well, and so a is a non-negative number.

The probability P that a given susceptible individual will not get infected in a unit time interval $(t, t + 1)$ then depends on the total number of infected persons at the time t — the larger that number is, the less will be the probability P. If $N(t) = 1$, then $P = q$; if $N(t) = 2$, then $P = q^2$; in general, $P = q^{N(t)}$. Then the probability for one healthy individual to get infected in the unit time interval following t is $1 - P = 1 - q^{N(t)}$, and the probable number of new cases of the disease will be proportional to the total number of healthy persons, that is, will be given by

$$V(t)(1 - q^{N(t)}) = V(t)(1 - e^{-aN(t)}).$$

What are the values of the functions $V(t)$ and $N(t)$ at $t + 1$? First of all, we have

$$N(t + 1) = V(t)(1 - e^{-aN(t)}) \qquad (4.12)$$

owing to the assumption that those who were ill at the time t will be well again after a unit time period, that is, at the time $t + 1$. On the other hand, for every t we evidently have

$$V(t) = X - N(t). \qquad (4.13)$$

For the sake of simplification, we pass now to relative values (we did the same in previous models). Put

$$x_1(t) = \frac{N(t)}{X}, \quad x_2(t) = \frac{V(t)}{X}, \quad aX = a.$$

Substituting in (4.12) and (4.13) we get

$$x_1(t + 1) = x_2(t)(1 - e^{-ax_1(t)})$$
$$x_2(t) = 1 - x_1(t)$$

or, after eliminating x_2 from the first equality,

$$x_1(t + 1) = (1 - x_1(t))(1 - e^{-ax_1(t)})$$

that is,

$$z_{n+1} = (1 - z_n)(1 - e^{-az_n}) \qquad (4.14)$$

where z_n is the relative number of sick individuals at time n. Equation (4.14) represents the desired model. Before analysing it, let us note some other models.

Example 4. Model (4.14) may be slightly modified. Suppose that an infected person is contagious for a unit period of time, the next unit period of time he or she is isolated (or immune — as in the case of influenza) and then he or she becomes susceptible again. Introducing, besides $N(t)$ and $V(t)$ from the previous example, the symbol $I(t)$ for the number of those who are immune at the time t, we get the following system of equations:

$$N(t + 1) = V(t)(1 - e^{-aN(t)})$$

$$I(t + 1) = N(t)$$

$$V(t) + I(t) + N(t) = X.$$

Introducing relative values again, with $x_3(t) = I(t)/X$, we obtain

$$x_1(t + 1) = x_2(t)(1 - e^{-ax_1(t)})$$

$$x_3(t + 1) = x_1(t)$$

$$x_1(t) + x_2(t) + x_3(t) = 1.$$

By eliminating x_3, the second and third equations yield

$$x_1(t) + x_2(t) + x_1(t - 1) = 1$$

and the first equation becomes

$$x_1(t + 1) = (1 - x_1(t) - x_1(t - 1))(1 - e^{-ax_1(t)})$$

or

$$z_{n+1} = (1 - z_n - z_{n-1})(1 - e^{-az_n}) \qquad (4.15)$$

where z_n denotes again the relative number of sick persons at the time n. Observe that (4.15) involves a function of two variables.

Example 5. With many diseases, the immunity period lasts longer than the disease itself. Assuming one time unit for the duration of the disease and two for the period of immunity (or isolation), we obtain the system

$$N(t + 1) = V(t)(1 - e^{-aN(t)})$$

$$I(t + 1) = N(t) + N(t - 1)$$

$$I(t) + N(t) + V(t) = X$$

which, analogously to the above case, reduces to

$$z_{n+1} = (1 - z_n - z_{n-1} - z_{n-2})(1 - e^{-ax_n})$$

where z_n denotes the relative number of sick persons. Readers will certainly find no difficulty in generalising the model to the case when the period of immunity or isolation lasts k time units.

The design of a mathematical model is the first, usually the easier, step. It is more difficult to investigate whether the model corresponds to reality and whether it has the properties it should have. Therefore, let us return to our first epidemiological model (4.14). Its properties, or rather the properties of the function employed in it, are stated in the following theorem.

Theorem 1

Let $f_a(x) = (1 - x)(1 - e^{-ax})$. Then

1. For all parameter values $a \geqslant 0$, the function f_a maps the interval $[0, 1]$ into $[0, 1]$.

2. *If $0 \leqslant a \leqslant 1$, then f_a has a unique fixed point $\alpha = 0$ and every sequence generated by any point $x_0 \in [0, 1]$ converges to α.*

3. *If $a > 1$, then f_a has two fixed points, a repulsive point $\alpha = 0$ and an attractive one $\beta \in (0, 1/2)$, and every sequence generated by any $x_0 \in (0, 1)$ converges to β.*

The graph of f_a, for some parameter values a, is depicted in figure 4.2.

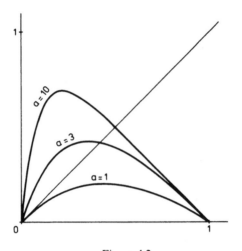

Figure 4.2

Theorem 1 allows us to interpret our model. If $a \leqslant 1$, then $\log q = -a/X \geqslant -1/X$, thus, if the coefficient q in our model is large enough (meaning little probability of getting infected), then the disease will completely disappear after some time. But if q falls under a certain limit, the disease will not disappear, but the percentage of sick individuals will stabilise after a sufficiently long time at a non-zero value of $100\,\beta$. This value is stable. This means that if the percentage of sick people is changed due to some temporary very unfavourable (or very favourable) circumstances, it will come back to that value after some time.

Theorem 1 can be proved as follows.

1. If $x \in [0, 1]$, then it also follows that $(1 - x) \in [0, 1]$ and $(1 - e^{-ax}) \in [0, 1]$. Therefore $f_a(x)$, being a product of two elements of $[0, 1]$, belongs to $[0, 1]$.

2. Evidently, $a = 0$ is a fixed point of f_a for all a. Suppose that $a \in [0, 1]$ and $x \in [0, 1]$, then $f_a(x) < x$. To verify it, examine the function

$$\varphi(x) = x - f_a(x) = 2x - 1 + e^{-ax} - x\,e^{-ax}.$$

Its derivative

$$\varphi'(x) = 2 + e^{-ax}[ax - (1 + a)]$$

is readily confirmed as positive, hence $\varphi(x)$ is an increasing function. Since $\varphi(0) = 0$, we get $\varphi(x) > 0$, that is $f_a(x) < x$. From the last inequality we easily obtain the assertion of the theorem.

3. Let $a > 1$. First of all, we have to verify that there is exactly one point $\gamma \in (0, 1)$ at which f_a attains its maximum. Further, we must check that in $(0, \gamma]$ the function is increasing and its graph lies above the line $y = x$ (i.e. $f_a(x) > x$), while in $[\gamma, 1]$ the function f_a is decreasing. The standard way of doing this is by examining the first and second derivatives of f_a. We omit that part of the proof. Readers acquainted with the method may do the proof as an exercise.

It follows from the properties stated above that f_a has exactly one fixed point, β, in the interval $(0, 1]$. The equality $f_a(\beta) = \beta$ reduces to

$$e^{-a\beta} = \frac{1 - 2\beta}{1 - \beta} > 0$$

and we get necessarily $0 < \beta < 1/2$.

To prove the remaining part of the theorem, we show that f_a has no 2-cycle. Then, by the Šarkovskii Theorem, it cannot have cycles of orders greater than 1. Therefore, according to Theorem 7 of the preceding chapter, the sequence generated by any $x_0 \in [0, 1]$ converges, the limit being a fixed point of f_a (from Theorem 5 of the same chapter). However, it is easy to see that $f_a'(0) > 1$, hence 0 is a repulsive fixed point. It can only be the limit for the sequences generated by $x_0 = 0$ or $x_0 = 1$, all other sequences converge to β.

Suppose, therefore, that f_a, for some $a > 1$, has a 2-cycle consisting of points x_1 and x_2. Let $x_1 < x_2$. Then

$$\frac{f_a(x_1) - f_a(x_2)}{x_2 - x_1} = \frac{x_2 - x_1}{x_2 - x_1} = 1.$$

The last equality yields

$$e^{-ax_1}(1 - x_1) = e^{-ax_2}(1 - x_2).$$

This, however, is impossible, as $e^{-ax_1} > e^{-ax_2} > 0$ (e^{-t} being a decreasing function) and $1 - x_1 > 1 - x_2 > 0$. The obtained contradiction shows that f_a has no 2-cycle, and the theorem is proved.

We are not going to analyse the remaining two models here. The form of the functions used in them makes it clear that it would not be an easy task. Readers should now have an idea, at least, of what problems are involved in mathematical modelling (and it was just the simplest models that we were studying). Another example of a more complicated model will be presented in the next section.

Exercises

4.5 Test the stability of fixed points of the equation

$$x_{n+1} = q(x_n) x_n$$

if

1. $q(x) = ax\, e^{1-ax}$ with $a > 0$,

*2. $q(x) = ax\, e^{-bx}$, where $a > b > 0$.

(Hint: in case 1 there is exactly one fixed point, in case 2 there are exactly two fixed points.)

4.6 Find the general solution of the difference equation

$$x_{n+2} = 4x_{n+1} + 5x_n$$

in the form $x_n = c_1\lambda_1^n + c_2\lambda_2^n$, where λ_i are uniquely determined constants and c_i are arbitrary coefficients.

4.4 A DISCRETE POPULATION GROWTH MODEL FOR TWO SPECIES

The most widespread relation between two biological species is that of a predator and its prey (i.e. the latter serves as food for the former). If we neglect all the non-homogeneities, both in the population (age, genetic structure) and in the environment (natural conditions), we can build a rather simple model for the evolution of a predator—prey population.

First, the notation. Denote by x_k the number of prey and by y_k the number of predators in a given territory at time k. Suppose that the reproduction of the prey obeys the model of equation (4.9) (this model is already rather general and yet not too complicated, therefore it is frequently used). Therefore,

$$x_{k+1} = ax_k - bx_k^2$$

where $a > 1, b > 0$. Next, assume that the number of predators is directly proportional to the number of prey and to the number of predators at a unit period of time before, i.e.

$$y_{k+1} = dx_ky_k$$

where $d > 0$ is the proportionality constant. Assuming that $cx_k y_k$ animals become victims of the predators, the number of prey at time $k + 1$ is given by

$$x_{k+1} = ax_k - bx_k^2 - cx_k y_k$$

where $c > 0$. In order to simplify the calculations, we introduce a new notation for the constants:

$$x_{k+1} = (1 + A)x_k - BDx_k^2 - CDx_k y_k \qquad (4.16)$$

$$y_{k+1} = Dx_k y_k$$

where A, B, C and D are positive numbers. Thus, we have just obtained a growth model for a predator—prey population.

We begin the analysis of the model, similarly as in preceding cases, by finding the fixed points. In the present case, we have to solve the following system of two non-linear equations with two unknowns x and y:

$$x = (1 + A)x - BDx^2 - CDxy$$

$$y = Dxy.$$

In the case where $y = 0$ there are no predators and the system reduces to the single equation

$$x = (1 + A)x - BDx^2$$

which we have already studied (see Exercise 4.3). We know that in this case also there exist two fixed points $x^1 = 0$ and $x^2 = A/(BD)$. If $y \neq 0$, then the fixed point has coordinates $\bar{x} = 1/D$, $\bar{y} = (A - B)/(CD)$, where $A > B$ owing to $y > 0$. Thus, the above system has three fixed points, described by the ordered pairs $[x^1, 0]$, $[x^2, 0]$ and $[\bar{x}, \bar{y}]$. A more detailed discussion of their properties exceeds the scope of this book. Readers may find detailed information in [11].

Readers interested in other possible applications are referred to [7], [21] and [26], where the problems mentioned in this chapter are studied in more detail, and additional bibliographical information is given.

5. FUNCTIONAL EQUATIONS IN ONE VARIABLE

5.1 INTRODUCTION

We have already encountered functional equations in Chapter 2. The methods mentioned there, however, are not suited to solve equations such as

$$f^2(t) = 0 \tag{5.1}$$

$$f^2(t) = t \tag{5.2}$$

$$f(3t) = f(t) + t^2 \tag{5.3}$$

$$f(t^2 - 2f(t)) = (f(t))^2 \tag{5.4}$$

involving a single independent variable t (f is the unknown function). The present chapter shows readers how such equations are solved.

It can be said, in general, that solving functional equations of one variable is much more difficult than solving functional equations in several variables, the solutions of the latter being, as a rule, either very 'neat' (regular) or very 'ugly' (strongly irregular). Moreover, all the continuous solutions of an equation in several variables usually form a parametrised system of functions depending on one or several constants (parameters). A typical example is provided by the Cauchy functional equation — all its continuous solutions form a parametrised system $\{f_a(t) = at;\ a \in R\}$, with one parameter a.

In the case of functional equations in one variable, the solutions are usuaiiy neither very neat nor very ugly. Besides, all the neat solutions of such functions often form a system parametrised by a function rather than a constant. We shall see this later in some examples.

As in Chapter 2, we do not give an exact definition of a functional equation in one variable — it is not an easy thing to do. Intuitively, readers certainly understand what is meant by that idea. Moreover, we shall mostly deal only with equations of the special type

$$f(g(t)) = G(t, f(t)) \qquad (5.5)$$

or

$$f(h(t, f(t))) = H(t, f(t)) \qquad (5.6)$$

where $g(x)$, $G(x, y)$, $h(x, y)$ and $H(x, y)$ are given functions (of one or two real variables) taking values in R, and $f(t)$ is the unknown function to be found. Its domain is always some set $A \subset R$. If the domain is not given, we try to solve the equation in the largest possible set $A \subset R$.

Note, as we shall also deal with iterates of functions in this chapter, the symbols $f^2, ..., f^n$ will denote the second up to the nth iterate of a function f, respectively, while $(f(x))^2, ..., (f(x))^n$ will denote its second up to its nth power. Thus, $f^2(x) = f(f(x))$ and $(f(x))^2 = f(x)f(x)$. In the discussion below we shall often make reference to some concepts and results introduced in Chapter 3.

Let us now consider the equations listed at the beginning of this chapter.

Example 1. Let us solve equation (5.1)

$$f^2(t) = 0$$

in an interval $I = [a, b]$, where $a \leqslant 0 \leqslant b$. Let f be the solution. Put $D_1 = \{t \in I; f(t) = 0\}$ and $D_2 = \{t \in I; f(t) \neq 0\}$. Evidently, $D_1 \neq \varnothing$ and $f(D_2) \subset D_1$. Also, $0 \in D_1$ (otherwise we would have $f(t) = 0 \in D_2$, and hence $f^2(t) \neq 0$, for any $t \in D_1 \neq \emptyset$).

The above three conditions ($D_1 \neq \emptyset, f(D_2) \subset D_1$ and $0 \in D_1$) are also sufficient for f to be a solution of the equation (5.1). In fact, if D_1 and D_2 are arbitrary disjoint sets satisfying $D_1 \cup D_2 = I$ and $0 \in D_1$, and if g is any mapping of D_2 into D_1, then the mapping f defined by

$$f(t) = \begin{cases} 0 & \text{if } t \in D_1 \\ g(t) & \text{if } t \in D_2 \end{cases}$$

solves equation (5.1). This solution parametrically depends on the function g (and on the set D_1).

What do all the continuous solutions of (5.1) look like? Let $J = f(I)$. As we assume that f is a continuous solution, J is necessarily an interval with end points $c \leq 0$ and $d \geq 0$. Then $f(J) = 0$, and hence $f(t) = 0$ for each $t \in [c, d]$ (the conditions $f(c) = f(d) = 0$ are evidently implied by continuity).

It is easy to verify that every continuous function f mapping I into I and satisfying $f(t) = 0$ for $t \in [c, d] \subset I$ and $f(t) \in [c, d]$ for all $t \in I$ is a solution of (5.1) (see figure 5.1).

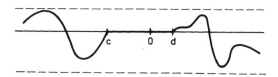

Figure 5.1

Example 2. Solve equation (5.2)

$$f^2(t) = t$$

in an interval $I = (a, b)$. Evidently, each point $t \in I$ is a fixed point of f^2 and two cases may arise: either $f(t) = t$ or

$f(t) \neq t$. In the latter case, the points t and $f(t)$ constitute a 2-cycle of f. Let

$$P = \{t \in I; f(t) = t\}$$
$$D_1 = \{t \in I; f(t) < t\}$$
$$D_2 = \{t \in I; f(t) > t\}.$$

Then obviously the sets P, D_1 and D_2 are pairwise disjoint and their union is the whole set I. Moreover, $f(D_1) = D_2$ and $f(D_2) = D_1$. The mapping f is at the same time injective and maps I onto I (why?). If f_1 is the restriction of f to D_1, then f_1 is an injective mapping of D_1 onto D_2 and, analogously, the restriction f_2 is an injective mapping of D_2 onto D_1. It is easy to check that $f_2 = f_1^{-1}$. The said properties determine, in their turn, any solution of (5.2). Whenever P, D_1 and D_2 are pairwise disjoint sets with $P \cup D_1 \cup D_2 = I$ and g is any injective mapping of D_1 onto D_2, then the function f mapping I onto I and defined by

$$f(x) = \begin{cases} x & \text{for} \quad x \in P \\ g(x) & \text{for} \quad x \in D_1 \\ g^{-1}(x) & \text{for} \quad x \in D_2 \end{cases}$$

solves equation (5.2). We see that this solution also depends parametrically on a function, namely the function g.

What are the continuous solutions of (5.2)? It is easy to find two such solutions. In the case $I = [-1, 1]$, they are

$$f_1(x) = x \qquad \text{and} \qquad f_2(x) = -x$$

and a more profound analysis discovers an infinity of such solutions. All of them have a common property: the graph of each of them is symmetrical with respect to the axis $y = x$ (why?). It can also be shown that the only continuous solution of our equation in I to have more than one fixed point is the identity function f_1.

In the above cases, we were solving rather simple equations. We managed to find all the solutions and we did so without using a complicated process. However, this is not always the case — the above examples are atypical. In more difficult cases we are happy to find just some solutions. The methods for solving such equations will be treated in the next section.

Exercises

5.1 Carry out the following.

1. Find all the solutions of the functional equation $f^3(t) = t$.
2. Can you find a continuous solution of this equation, except the identity function $f(t) = t$?

5.2 Can you find a continuous solution of $f^4(t) = t$ that would not be a solution of the equation $f^2(t) = t$?

5.3 1. Find all the solutions of $f^3(t) = 0$ in an interval I.
2. What do the continuous solutions of this equation look like?

5.4 Show that every continuous solution of equation (5.2) in an interval I, different from the identity function, has exactly one fixed point.

5.2 CHARACTERISTIC MAPPINGS AND INVARIANTS

Notice that whenever f is a solution of (5.5) defined on a set A, then for each $t \in A$ we have $g(t) \in A$ as well. Further, let Γ_f be the graph of a solution f of (5.5). For any $t_0 \in A$, put $x_0 = f(t_0)$. Then the ordered pair (i.e. the point in the plane) $[t_0, x_0]$ belongs to Γ_f. Now if $t_1 = g(t_0)$ and $x_1 = G(t_0, x_0)$, then also $[t_1, x_1] \in \Gamma_f$. This implies that the mapping

$$\chi: [t, x] \mapsto [g(t), G(t, x)]$$

assigning to $[t, x]$ the pair $[g(t), G(t, x)]$ maps the graph Γ_f into itself. (The domain and the range of χ are all the plane.) The mapping χ is termed the characteristic mapping of the functional equation (5.5). It allows us, in some cases, to find at least some solutions. Let us show how.

Example 3. Solve equation (5.3)

$$f(3t) = f(t) + t^2$$

in R. Its characteristic mapping χ is defined by $t \mapsto 3t$, $x \mapsto x + t^2$. Put $\varphi(t, x) = x - t^2/8$. We readily see that $\varphi(3t, x + t^2) = x + t^2 - 9t^2/8 = \varphi(t, x)$. Thus the value of the function $\varphi(t, x)$ is left unchanged by χ. Let $c \in R$ and let B_c denote the set of all the points in the plane satisfying $\varphi(t, x) = c$, that is $x - t^2/8 = c$. Then the set B_c is mapped into itself by χ. At the same time, B_c is the graph of the function $f(t) = t^2/8 + c$, which is easily confirmed to be a solution of (5.3). However, it is not the only solution, but that we shall see later.

The above method can also be used for equations of the type (5.6). In that case, the characteristic mapping takes the form

$$\chi: [t, x] \mapsto [h(t, x), H(t, x)].$$

Example 4. Using the characteristic mapping, solve equation (5.2) in R. We have $\chi: [t, x] \mapsto [x, t]$. We seek a suitable function $\varphi(t, x)$ whose values would not be changed by the transformation χ. For instance, we may choose $\varphi(t, x) = tx$. Let $B_c = \{[t, x]; \varphi(t, x) = c\}$. Then B_c is the graph of the function $f(t) = c/t$, defined for all $t \neq 0$. Putting $f(0) = 0$, we get a solution of (5.2).

Instead of φ we may take another function, say $\psi(t, x) = t + x$. Analogously we obtain the solution $f(t) = -t$ or, more generally, $f(t) = -t + c$ with c a constant. Another convenient function is, for example, $\tau(t, x) = (t - x)^2$ (verify it). The equality $(t - x)^2 = 0$ yields a solution of (5.2). The equality $(t - x)^2 = c$ gives two functions $f(t) = t \pm \sqrt{c} = t \pm d$, where d is a constant, but neither of them is a solution for $d \neq 0$. However, if we interpret $f(t) = t \pm d$ in such a way that the sign of d depends on t, that is, if we put $f(t) = t + dg(t)$, where g is a suitable function of t with values $+1$ and -1, then we may obtain a solution (try to find it).

The auxiliary functions φ, ψ and τ in Examples 3 and 4 are invariants of the respective mappings χ in the sense of the following definition.

Definition 1. Let χ be a mapping of the plane $R \times R$ into $R \times R$ and let $\varphi(u, v) = \varphi([u, v])$ be a function of two variables.

1. If for every $[u, v] \in R \times R$ the equation

$$\varphi([u, v]) = \varphi(\chi([u, v]))$$

holds, then φ is termed an invariant of χ.

2. Analogously, if $A \in R \times R$ is any set satisfying

$$\chi(A) \subset A$$

then A is said to be an invariant set with respect to the mapping χ.

It is easy to see that if φ is an invariant and c is a number, then

$$A_c = \{[u, v] \in R \times R; \ \varphi(u, v) = c\}$$

is an invariant set. Of course, it may happen that $A_c = \varnothing$ or $A_c = R \times R$. We are not interested in these cases but in the invariants that determine some functions or, equivalently, in invariant sets which are graphs of some functions. Searching for such invariants is an art in itself. There exists a theory that may facilitate the search, but it is so com-

-plicated that it exceeds the scope of this book. Therefore, we give here a few guidelines only.

First of all, whenever $\varphi_1(u, v)$ and $\varphi_2(u, v)$ are invariants of the characteristic mapping of some functional equation and S is an arbitrary function of two variables, then $S(\varphi_1(u, v), \varphi_2(u, v))$ is an invariant as well. Also, if A_1 and A_2 are invariant sets (for the same mapping), then their intersection too is an invariant set. This can be generalised for any finite number of invariants or invariant sets.

If the characteristic mapping χ is periodic, i.e. if χ^k is the identity mapping of $R \times R$ for some positive integer k, then it is easy to find an invariant. Let $\varphi(u, v)$ be an arbitrary function and put $\varphi_i(u, v) = \varphi(\chi^i([u, v]))$. It is not difficult to verify that

$$\Phi(u, v) = \varphi_1(u, v)\, \varphi_2(u, v) \ldots \varphi_k(u, v)$$

is an invariant of χ. Let us illustrate this by an example.

Example 5. Solve the equation

$$f(a - t) = b - f(t) \tag{5.7}$$

in R, where a and b are constants. Its characteristic mapping is defined by $\chi: [t, x] \mapsto [a - t, b - x]$ and the second iterate χ^2 of the characteristic mapping is the identity mapping. Let $\varphi(u, v) = u$. Then

$$\tilde{\varphi}(t, x) = \varphi(t, x)\, \varphi(\chi([t, x])) = t(a - t)$$

is an invariant of χ. Similarly, by putting $\psi(u, v) = v$, we obtain the invariant

$$\bar{\psi}(t, x) = x(b - x).$$

If Φ is an arbitrary function of two variables, then

$$\Phi(at - t^2,\, bx - x^2)$$

is also an invariant. Using a particular choice of Φ (putting

$\Phi(u, v) = -v + cu = 0$) we get the equation

$$x^2 - bx + c(at - t^2) = 0 \qquad (5.8)$$

where c is an arbitrary constant. By solving the quadratic equation (5.8) with unknown x we get

$$x = \frac{b}{2} \pm \left(\frac{b^2 - 4c(at - t^2)}{4} \right)^{1/2}.$$

Choosing a suitable constant c to simplify the square root, say $c = b^2/a^2$, yields two solutions

$$f(t) = \frac{b}{2} \pm b \left(\frac{t}{a} - \frac{1}{2} \right)$$

of our equation. Again, other solutions also exist.

Example 6. A functional equation having some invariants need not have a solution. The equation

$$f(-t + 1) = f(t) + 1 \qquad (5.9)$$

has invariants $t - t^2$ and $\varphi(x)$, where φ is any periodic function with period 1 (verify this). However, if (5.9) had a solution defined at least at two points, t_0 and $1 - t_0$, then we would have

$$f(1 - t_0) = f(t_0) + 1$$

and, at the same time

$$f(1 - (1 - t_0)) = f(1 - t_0) + 1.$$

That is,

$$f(1 - t_0) = f(t_0) - 1$$

which is impossible. Therefore, (5.9) cannot have a solution in any set $A \neq \varnothing$.

Example 7. Now consider the equation

$$f(t^2 - 2f(t)) = (f(t))^2. \qquad (5.10)$$

Its characteristic mapping χ is defined by

$$\chi: [t, x] \mapsto [t^2 - 2x, x^2].$$

Examine the function $\varphi(t, x) = t^2/x$. We can easily verify that

$$\varphi(\chi([t, x])) = \left(\frac{t^2}{x} - 2\right)^2 = (\varphi(t, x) - 2)^2.$$

Hence, $\varphi(t, x)$ is not an invariant, yet the mapping $g(z) = (z - 2)^2$ has two fixed points, $z_1 = 1$ and $z_2 = 4$. Therefore,

$$A_1 = \{[t, x];\ \varphi(t, x) = 1\}$$

is an invariant set for χ, as well as the set

$$A_2 = \{[t, x];\ \varphi(t, x) = 4\}.$$

On the other hand, A_1 is the graph of the function $f(t) = t^2$ and A_2 the graph of $f(t) = t^2/4$. Both functions are solutions of (5.10).

The function g has a 2-cycle (verify it). Its points α_1 and α_2 determine another invariant set

$$A_3 = \{[t, x];\ (\varphi(t, x) - \alpha_1)(\varphi(t, x) - \alpha_2) = 0\}.$$

In fact, if $[t, x] \in A_3$, then necessarily $\varphi(t, x)$ must equal one of the numbers α_1 or α_2. Suppose, for instance, that $\varphi(t, x) = \alpha_1$ (in the other case we proceed analogously). Then

$$\varphi(\chi([t, x])) - \alpha_2 = (\varphi(t, x) - 2)^2 - \alpha_2 = (\alpha_1 - 2)^2 - \alpha_2 = 0$$

meaning that $\chi([t, x]) \in A_3$. If we require that the function determined by A_3 be continuous, we obtain (after checking the two possible equation solutions)

$$f_1(t) = \begin{cases} t^2/\alpha_1 & \text{for } t \geq 0 \\ t^2/\alpha_2 & \text{for } t < 0 \end{cases} \qquad f_2(t) = \begin{cases} t^2/\alpha_2 & \text{for } t \geq 0 \\ t^2/\alpha_1 & \text{for } t < 0. \end{cases}$$

Exercises

5.5 Using the method of invariants, solve the following functional equations (a and b are constants):

1. $f(t + f(t)) = 3f(t)$;
2. $f(f(t) - t) = f(t)$;
3. $f(f(t) - 2t) = 2f(t) - 3t$;
4. $f(t + a) = f(t) + b$;
5. $f(t + a) = -f(t) + b$.

*5.6 Find explicitly the solution of (5.10) corresponding to the 2-cycle of the function $g(z) = (z - 2)^2$ mentioned in Example 7 and check that it really is a solution.

5.7 The idea of an invariant can also be introduced in the case of a mapping, χ, of A into A, where A is any subset of the plane (i.e. the Cartesian product) $R \times R$; such invariants and the deduced invariant sets may even be used for solving equations.

1. Formulate an exact definition of such an invariant.
2. Solve the equation below by the modified method:

$$f(\sqrt{t} - f(t)) = 2f(t) + t - \sqrt{t}$$

5.3 LINEAR FUNCTIONAL EQUATIONS

Linear functional equations are those of the type

$$f(g(t)) = h(t)f(t) + F(t) \tag{5.11}$$

or, in the particular case with $F(t)$ identically equal to zero,

$$f(g(t)) = h(t)f(t). \tag{5.12}$$

Here, $g(t)$, $h(t)$ and $F(t)$ are given functions and the function $f(t)$ is the unknown; just as before we shall assume that all the functions are defined on some interval I and that $g(I) \subset I$.

Linear equations have a privileged position in the theory of functional equations because, as a rule, they are easier to solve than other equations. An equation of the type (5.11) is termed a non-homogeneous functional equation, while (5.12) is said to be homogeneous. When solving equations of the type (5.11), the following theorem often proves useful, as it allows us to reduce, under certain conditions, the problem of solving the more complicated equation (5.11) to solving the simpler equation (5.12).

Theorem 1

Let f_0 be any solution of (5.11). *Then*

1. *whenever f_1 is a solution of* (5.12), *the function*

$$f(t) = f_0(t) + f_1(t)$$

solves the equation (5.11);

2. *every solution of* (5.11) *can be obtained from f_0 in the way stated in part* 1.

The proof is simple. Part 1 is readily verified by substitution. To prove part 2, denote by f an arbitrary solution of (5.11). Then $f_1(t) = f(t) - f_0(t)$ solves (5.12), and hence the function $f(t) = f_0(t) + f_1(t)$ has the required form.

What Theorem 1 really states is that all the solutions of a non-homogeneous equation are obtained by adding all the solutions of the corresponding homogeneous equation to a chosen (particular) solution of the non-homogeneous equation.

The following theorem concerns the solutions of a homogeneous equation.

Theorem 2

1. *The sum of any two solutions of the homogeneous equation* (5.12) *is again a solution of* (5.12).

2. *Let f be a solution of* (5.12). *If $\varphi(t)$ is an invariant of its characteristic mapping depending only on the first variable t (in particular, if $\varphi(t)$ is a constant) and if $w(t)$ is any function, then*

$$f_1(t) = w(\varphi(t))f(t)$$

is a solution of (5.12).

To prove this, the assertion in part 1 is easily shown to be true by substitution. Part 2 follows from $w(\varphi(g(t))) = = w(\varphi(t))$, implying by (5.12) that

$$w(\varphi(g(t)))f(g(t)) = w(\varphi(t))h(t)f(t)$$

which was to be proved.

Example 8. Let us have another look at equation (5.7),

$$f(a - t) = b - f(t)$$

whose two solutions were found in Example 5. One of them is

$$f_0(t) = \frac{b}{2} + b\left(\frac{t}{a} - \frac{1}{2}\right).$$

From Theorem 1, it is now sufficient to find the solutions of the homogeneous equation

$$f(a - t) = -f(t). \tag{5.13}$$

The characteristic mapping is given by $\chi: [t, x] \mapsto \mapsto [a - t, -x]$. Consider the functions

$$\varphi_1(t, x) = t - \frac{a}{2} \qquad \varphi_2(t, x) = x.$$

We readily observe that $\varphi_1(\chi([t, x])) = -\varphi_1(t, x)$ and $\varphi_2(\chi([t, x])) = -\varphi_2(t, x)$, hence

$$\psi(t, x) = \varphi_1(t, x)/\varphi_2(t, x) = \left(t - \frac{a}{2}\right)x^{-1}$$

is an invariant, and as a possible solution of (5.13) we get

$$f_1(t) = c\left(t - \frac{a}{2}\right).$$

On the other hand, we know that $t(a - t)$ is an invariant (see Example 5). Therefore, Theorem 2 implies that if w_1 is any function, then

$$f_2(t) = f_1(t) w_1(at - t^2)$$

also solves equation (5.13). It follows from Theorem 1 that $f = f_0 + f_2$ is a solution of (5.7). It reduces to

$$f(t) = \frac{b}{2} + \left(t - \frac{a}{2}\right)w(at - t^2) \qquad (5.14)$$

(where w is an arbitrary function). It can be shown that (5.7) has no other solutions; equation (5.14) is the general solution of (5.7).

Example 9. The equation

$$f(t + a) = f(t)$$

has the characteristic mapping $\chi: t \mapsto t + a, \ x \mapsto x$. This gives $t/a \mapsto t/a + 1$. If w is any periodic function with period 1 (that is $w(t + 1) = w(t)$ for all t), then $w(t/a)$ is an invariant of χ, as well as a solution of the above equation. It can be shown to be the general solution of that equation (i.e. the equation has no other solutions). The general solution of the equation

$$f(t + a) = f(t) + b \qquad (5.15)$$

(see Exercise 5.5 part 4) is obtained as the sum of some particular solution of (5.15), say $f_0(t) = bt/a$, and the function $w(t/a)$, that is

$$f(t) = \frac{b}{a}t + w\left(\frac{t}{a}\right).$$

In the linear equations we have solved so far, the function $h(t)$ was always constant. What really poses a problem, however, is to solve equations of the type (5.11) or (5.12) in which $h(t)$ is not a constant function. We present here a method that in some cases makes it possible to simplify a homogeneous equation with a non-constant coefficient $h(t)$.

Consider the homogeneous equation (5.12)

$$f(g(t)) = h(t)f(t)$$

in which $g(0) = 0$ and $h(0) = \lambda \neq 0$. After the substitution

$$f(t) = y(t)/z(t)$$

and an evident reduction, we obtain the equation

$$y(g(t))z(t) = \lambda y(t)\frac{h(t)}{\lambda}z(g(t))$$

which decomposes into two simpler equations

$$y(g(t)) = \lambda y(t) \quad \text{and} \quad z(t) = \frac{h(t)}{\lambda}z(g(t)).$$

Now the first of these equations has a constant coefficient λ, and the coefficient $h(t)/\lambda$ of the second equation has the value 1 at $t = 0$, which is sometimes advantageous. Here is an example.

Example 10. Let us solve the equation

$$f(at) = \lambda\, e^{ht}f(t) \qquad (5.16)$$

where a, b and λ are constants, $a > 0$ and $\lambda > 0$. The substitution $f(t) = y(t)/z(t)$ yields two equations

$$y(at) = \lambda y(t) \tag{5.17}$$

and

$$z(at)\, e^{bt} = z(t). \tag{5.18}$$

These equations being rather complicated, we shall try to find the invariants by the following method, which is not completely correct, but often provides a result. Of course, we shall have to check anyway that what we have obtained really is an invariant.

Solve equation (5.17) first. We write χ in terms of coordinates, that is

$$t \mapsto at, \qquad x \mapsto \lambda x$$

and apply suitable functions to both sides of these relations until we get something reasonable. If we restrict ourselves to $t \geqslant 0$ and apply the logarithmic function, we get

$$\ln t \mapsto \ln t + \ln a, \qquad \ln x \mapsto \ln x + \ln \lambda$$

and, hence, using the functions $y/\ln a$ and $y/\ln \lambda$, we obtain

$$\frac{\ln t}{\ln a} \mapsto \frac{\ln t}{\ln a} + 1 \quad \text{and} \quad \frac{\ln x}{\ln \lambda} \mapsto \frac{\ln x}{\ln \lambda} + 1.$$

(Now, of course, the arrow does not mean the mapping χ nor any of its components, but another mapping. Do you know which one?) Thus we obtain the invariants

$$\frac{\ln t}{\ln a} - \frac{\ln x}{\ln \lambda} \quad \text{and} \quad \omega\!\left(\frac{\ln t}{\ln a}\right)$$

where ω is an arbitrary periodic function with period 1 (verify this). Letting the first invariant equal zero, we compute one solution which, combined with the second invariant using Theorem 2, yields the solution

$$y(t) = t^{\ln \lambda / \ln a} \, \omega(\ln t / \ln a)$$

for equation (5.17).

Now solve, in a similar way, equation (5.18). The characteristic mapping is defined by

$$\chi: t \mapsto at, \qquad x \mapsto e^{-bt}x$$

which reduces to

$$t \mapsto at, \qquad \ln x \mapsto -bt + \ln x.$$

Try to find an invariant in the form $\ln x + ct$, with c a suitable constant. After the substitution, we compute $c = = b/(1 - a)$. Therefore

$$\ln z(t) = \frac{b}{1 - a} t$$

that is

$$z(t) = e^{bt/(1 - a)}$$

is a solution of (5.18). The desired solution of the original equation (5.16) is then

$$f(t) = \frac{y(t)}{z(t)} = t^{\ln \lambda / \ln a} \, e^{bt/(1 - a)} \, w\!\left(\frac{\ln t}{\ln a}\right).$$

We conclude this section with a general theorem on the existence of continuous solutions of linear functional equations. But first we must introduce the necessary notation. Let I be any interval and let ξ be its end point (it may also be $\xi = +\infty$ or $\xi = -\infty$). Suppose that g is an increasing and continuous function on the interval I with $g(I) \subset I$ and that the following conditions are met: if ξ is the left end point of I, then $g(t) < t$ for all $t \in I$; if ξ is the right end point, then $g(t) > t$ for all $t \in I$. Then we say that the function $g(t)$ belongs to the class $R_\xi(I)$. One such function is shown in figure 5.2; its graph must lie in the shaded region. Evidently, if $g \in R_\xi(I)$, then for each $t \in I$ we have $\lim_{n \to \infty} g^n(t) = \xi$.

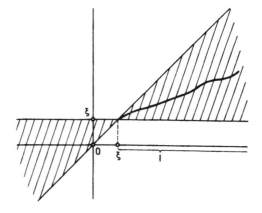

Figure 5.2

Now we can formulate the theorem below, for equation (5.11).

Theorem 3

Suppose that g is a function in the class $R_\xi(I)$ with $\xi \notin I$. Further, let h and F be continuous functions defined on I and let $h(t) \neq 0$ for all $t \in I$. Then the equation

$$f(g(t)) = h(t)f(t) + F(t)$$

has a continuous solution in the interval I (see Kordylewski and Kuczma [15]).

More precisely, given any $t_0 \in I$ and any continuous function $\varphi_0(t)$ defined on a closed interval I_0 with end points t_0 and $g(t_0)$ (that is $I_0 = [t_0, g(t_0)]$ or $I_0 = [g(t_0), t_0]$) and satisfying

$$\varphi_0(g(t_0)) = h(t_0)\,\varphi_0(t_0) + F(t_0) \qquad (5.19)$$

there exists exactly one solution f of (5.11) which coincides on I_0 with φ_0, that is

$$f(t) = \varphi_0(t) \quad \text{for} \quad t \in I_0. \tag{5.20}$$

This solution is continuous.

In short, the theorem states that, under the mentioned assumptions, any continuous function defined on I_0 and satisfying (5.19) can be continuously extended to the whole interval I so that (5.11) holds for all $t \in I$. The function φ_0 is therefore referred to as the initial function.

The proof is as follows. For the sake of simplification, we shall suppose that ξ is the left end point (in the other case, the proof is similar). Let φ_0 be a continuous function defined on the interval $I_0 = [g(t_0), t_0]$ and satisfying (5.19).

The solution of the equation will be obtained by progressively extending the function φ_0. In the first step we may extend its domain to the interval $I_1 = [g^2(t_0), g(t_0)]$. In fact, if f is a solution of (5.11) coinciding with φ_0 on I_0, then

$$f(g(t)) = h(t)\,\varphi_0(t) + F(t) \tag{5.21}$$

for all $t \in I_0$. Since for $t \in I_0$ we have $g(t) \in I_1$, equation (5.21) defines unambiguously the values of f in I_1, and does so on the whole of I_1 because g is an injective mapping of I_0 onto I_1. Also, f is continuous on I_1. To prove it, choose an arbitrary $s \in I_1$. As $s = g(t)$ for some $t \in I_0$, we have $t = g^{-1}(s)$, where g^{-1} is the inverse function of g. Then (5.21) takes the form

$$f(s) = h(g^{-1}(s))\,\varphi_0(g^{-1}(s)) + F(g^{-1}(s)).$$

The function on the right-hand side, being constructed in an admissible way from continuous functions (we know that g^{-1} is continuous owing to g being continuous and increasing), is continuous, and hence $f(s)$ is continuous for $s \in I_1$.

We have not yet made use of (5.19). It is useful because it prevents the function f, defined already on $[g^2(t_0), t_0]$, from having two values at $g(t_0)$, namely $\varphi_0(g(t_0))$ and at the same time $h(t_0)\,\varphi_0(t_0) + F(t_0)$ prescribed by (5.21) for $t = t_0$. Next, the values of f in the interval $I_2 = [g^3(t_0), g^2(t_0)]$ are determined using (5.21) by their values in I_1 etc.

In order to get to the right of I_0 as well, it suffices to write the equation (5.11) in the form

$$f(t) = \frac{f(g(t)) - F(t)}{h(t)} = \Phi(t)$$

(we are making use of the fact that $h(t) \neq 0$). For $t \in I_{-1} = [t_0, g^{-1}(t_0)]$ we have $g(t) \in I_0$, and therefore the function $\Phi(t)$ is defined by the known values of f in I_0 (and by values of the known functions g, h, F). The continuity of f on I_{-1} is shown in a similar way to the preceding cases. The next step gives the values of f in $I_{-2} = [g^{-1}(t_0), g^{-2}(t_0)]$ etc. Since every point of I lies in some I_n and f is continuous on I_n, the function is consequently continuous on the whole interval I. The proof of the theorem is complete.

Readers might have been surprised by the above proof, or perhaps they skipped it all. It is sufficient to know that a solution exists. Still, the proof given above is very sensible, it is a constructive proof. It provides a method for seeking the solution; in a concrete case, it is sufficient to proceed in just the same way as we did in the proof. Depending on how many steps we perform, we shall obtain a solution defined on a larger or smaller part of I. Of course, we would prefer a solution defined on the whole interval I, but if we cannot find such a solution the partial solution is quite valuable. Similar situations occur in mathematics very often.

Exercises

5.8 Solve the equations:

1. $x(2t) = \lambda\,(\cos t)\,x(t)$, where $\lambda > 0$;

2. $x(2t) = (\cos t) x(t) - (\sin t)^2$.

5.9 Solve the equation (5.16) from Example 10 for the case where a and λ are arbitrary and non-zero, but not necessarily positive, constants and $t \in R$.

5.10 Solve the equation $x(at) = t^m x(t)$, where $a > 0$, $a \neq 1$. (Hint: find the invariant in the form $\ln x + p(\ln t)$, where p is a suitable polynomial.)

5.11 By the method used in the proof of Theorem 3, find a solution of the functional equations:

1. $f(t^2) = (t^2 + 1)f(t) + t$ in the interval $[1/8, 1/\sqrt{2}]$ if $t_0 = 1/2$ and $\varphi_0(t) = 1 - 2t$;
2. $f(2') = (t^2 + 1)f(t) - t$ in the interval $(-\infty, 2]$, if $t_0 = 0$ and $\varphi_0(t) = t^2 - t$.

5.4 THE ABEL AND SCHRÖDER EQUATIONS

The following equations are also important for solving other functional equations: the Abel equation

$$f(g(t)) = f(t) + 1 \qquad (5.22)$$

and the Schröder equation

$$f(g(t)) = \lambda f(t) \qquad (5.23)$$

($\lambda \neq 0$ being a constant).

In the case of the first equation, a simple condition for the existence of a solution can be stated as follows.

Theorem 4

The equation (5.22) *has a solution if, and only if, g has neither cycles nor fixed points.*

In that case, equation (5.22) *has infinitely many solutions.*

To prove this we proceed as follows. It is easy to observe that for every t and each positive integer n we have

$$f(g^n(t)) = f(t) + n. \tag{5.24}$$

Hence, if g has a cycle of order $k \geqslant 1$ generated by some point y, then (5.24) implies

$$f(y) = f(g^k(y)) = f(y) + k \neq f(y)$$

which is not possible.

In order to prove the second part of the theorem, we introduce some further ideas. Any function g mapping a set A into A can be used to define an equivalence relation in A as follows: two elements $x, y \in A$ are said to be equivalent, denoted $x \sim_g y$, if, and only if, there exist positive integers m and n with $g^m(x) = g^n(y)$. It is easy to verify that it really is an equivalence relation. In fact, for any three elements $x, y, z \in A$ we have $x \sim_g x$ (reflexivity), $x \sim_g y$ implies $y \sim_g x$ (symmetry), and $x \sim_g y$ and $y \sim_g z$ together imply $x \sim_g z$ (transitivity). Now assign to every $x \in A$ the set $\Omega(x)$ (or more precisely $\Omega_g(x)$) of all elements equivalent to x. The set $\Omega(x)$ is termed the orbit of x. Just as with any equivalence relation, the orbits form a partition of the set A, which means that every $x \in A$ belongs to exactly one orbit. In other words, if x and y are elements of A, then either $\Omega(x) = \Omega(y)$ or $\Omega(x) \cap \Omega(y) = \varnothing$.

Now we can prove the second part of the theorem. Choose an element $x_0 \in R$ and put $f(x_0) = y_0$, where y_0 is any number. We are going to show that this already unambiguously defines the function everywhere on the orbit $\Omega(x_0)$. Employing (5.24), we may write

$$f(g^n(x_0)) = y_0 + n$$

for $n = 1, 2, \ldots$. Since the function g has neither fixed points nor cycles, the definition is correct. Further, if $z \in \Omega(x_0)$ and if k and s are the least positive integers with $g^k(z) = g^s(x_0)$, then

$$f(g^k(z)) = f(g^s(x_0)) = y_0 + s$$

and so

$$f(z) = y_0 + s - k.$$

Thus, f has been defined on the whole orbit $\Omega(x_0)$. In a similar manner we define the function f on other orbits. Evidently, if $\Omega(x_1)$ and $\Omega(x_2)$ are disjoint orbits, then f is defined on $\Omega(x_2)$ independently of how it was defined on $\Omega(x_1)$, because g maps every orbit $\Omega(x)$ into itself.

It is evident from the construction that f is a solution of (5.22). We note that there are infinitely many solutions corresponding to infinitely many ways in which y_0, that is the value of f at x_0, can be chosen. The theorem is proved. Notice that the theorem holds even if the domain of g is any, not necessarily numerical, set A with $g(A) \subset A$. As we have not assumed g to be continuous, we cannot speak about continuity of the solution f. However, the continuity of g is far from guaranteeing the existence of a continuous solution f. It is only clear that g must not have fixed points. That is why we assumed in Theorem 3, among other things, that f belongs to the class $R_\xi(I)$. Of course, (5.22) may also have a solution for other functions g. For the details, readers are referred to [16].

The Schröder equation is related to the Abel equation. If in (5.23) we have $\lambda > 0$, $\lambda \neq 1$ and if f is a solution of (5.22), then the function $\varphi(t) = \lambda^{f(t)}$ solves the Schröder equation (5.23). However, even for the case where $\lambda > 0$ and $\lambda \neq 1$, equation (5.23) may also have other solutions.

Now let us examine in more detail equation (5.23). It always has at least one solution — the zero constant. For establishing the existence of solutions in a neighbourhood of a fixed point ξ of g, the following condition is sometimes useful: if the functions f and g have derivatives at ξ and if (5.23) holds, then

$$f'(\xi)g'(\xi) = f'(g(\xi))g'(\xi) = \lambda f'(\xi)$$

i.e. either $f'(\xi) = 0$ or $g'(\xi) = \lambda$. The importance of both these equations lies in the possibility they offer of simplifying the solution of more complicated equations, in some cases at least.

If g is a non-linear function having a fixed point ξ, that is $g(\xi) = \xi$, and if we can find a continuous increasing or decreasing solution of the Schröder equation $\varphi(g(t)) = = \lambda\varphi(t)$, where $\lambda \neq 0, \pm 1$, then we may simplify the solution of equation (5.11)

$$f(g(t)) = h(t)f(t) + F(t).$$

Applying the substitution

$$f(t) = \psi(\varphi(t)), \quad \varphi(t) = \tau$$

we obtain a new equation

$$\psi(\lambda\tau) = h(\varphi^{-1}(\tau))\,\psi(\tau) + F(\varphi^{-1}(\tau))$$
$$= h^*(\tau)\,\psi(\tau) + F^*(\tau). \qquad (5.25)$$

This method is called linearisation. Usually, the equation (5.25) is easier to solve than the given equation (5.11).

Similarly, using monotone continuous solutions of the Abel equation, (5.11) reduces to the form

$$\psi(\tau + 1) = h^*(\tau)\,\psi(\tau) + F^*(\tau).$$

Conditions guaranteeing the existence of a monotone continuous solution of (5.22) or (5.23), or the existence of solutions having other properties, can be found in the book mentioned above [16].

We give one more example.

Example 11. Let us solve the equation

$$f(t^a) = \lambda t^b f(t)$$

where a, b and λ are constants ($a > 0, \lambda > 0$). We use the method of linearisation. The equation

$$\varphi(t^a) = a\varphi(t)$$

has a continuous increasing solution $\varphi(t) = \ln t$. By the substitution

$$f(t) = \psi(\ln t) \quad \tau = \ln t$$

we get a new equation

$$\psi(a\tau) = \lambda\, e^{b\tau}\psi(\tau)$$

which, however, is the equation from Example 10 that has already been solved. By the inverse substitution we then obtain a solution of the original equation.

Exercise

5.12 Use the linearisation method to solve the equation

$$f((t - a)^2 + a) = bf(t).$$

5.5 CONCLUDING REMARKS

What has been said in this chapter is just an introduction to the interesting and useful study of functional equations in one variable. The aim was to outline to readers the problems that can be solved in this region and to acquaint them with the most commonly used methods of this non-traditional mathematical discipline. It should be remarked that the theory of functional equations has been developing very rapidly in the past few years and an elementary book like this one cannot cover much of the progress made.

Therefore, we refer readers who have a deeper interest in the subject to the literature listed in the references. Worthy of special mention are the accessible books by Šarkovskii and Peljuch [36] and by an outstanding expert in this field, Polish mathematician M Kuczma [16].

REFERENCES

1. Aczél, J.: *Lectures on Functional Equations*. New York, Academic Press, 1966.
2. Aczél, J.—Dhombres, J.: *Functional Equations Containing Several Variables. Encyclopedia of Mathematics and its Applications*. Cambridge, Cambridge University Press, 1987.
3. Block, L.: Stability of periodic orbits in the theorem of Šarkovskii. *Proc. Amer. Math. Soc.* **81**, 1981, pp. 333 – 336.
4. Block, L.—Guckenheimer, J.—Misiuriewicz, M.—Young, L. S.: Periodic points and topological entropy of one dimensional maps. *Global Theory of Dynamical Systems. Lecture Notes in Mathematics,* **819**. Berlin, Springer 1980, pp. 18 – 34.
5. Burkart, U.: Interval mapping graphs and periodic points of continuous functions. *J. Comb. Theory* B, **32**, 1982, pp. 57 – 68.
5a. Campanino, M.—Epstein, H.: On the existence of Feigenbaum's fixed point. *Commun. Math. Phys., ***79**, 1981, pp. 261 – 302.
6. Collet, P.—Eckmann, J. P.: *Iterated Maps on the Interval as Dynamical systems. Progress in Physics Series*. Basel, Birkhäuser, 1980.
7. Cooke, K. L.—Calef, D. F.—Level, E. V.: Stability or chaos in discrete epidemic systems. *Proceedings of the Conference on Nonlinear Systems and Applications*. New York, Academic Press, 1977.
8. Dhombres, J.: *Some Aspects of Functional Equations*. Bangkok, Chulalongkorn University, 1979.
9. Engel, A.: Systematic use of applications in mathematics teaching. *Educ. Stud. Math.,* **1**, 1968 – 69, pp. 202 – 221.
10. Feigenbaum, M. J.: Quantitative universality for a class of nonlinear transformations. *J. Stat. Phys.* **19**, 1978, pp. 25 – 52.
11. Hadeler, K. P.: *Mathematik für Biologen*. Berlin, Springer, 1974.
12. Janková, K.—Smítal, J.: A characterization of chaos. *Bull. Austral. Math. Soc.,* **34**, 1986, pp. 283 – 292.

13. Kloeden, P. E.: Chaotic difference equations are dense. *Bull. Austral. Math. Soc.*, **15**, 1976, pp. 371 – 379.

14. Kominek, Z.: On the continuity of Q-convex functions and additive functions. *Aequationes Math.*, **23**, 1981, pp. 146 – 150.

15. Kordylewski, J.—Kuczma, M.: On some linear functional equations I, II. *Ann. Polon. Math.* **9**, 1960, pp. 119 – 136. **11**, 1962, pp. 203 – 207.

16. Kuczma, M.: *Functional Equations in a Single Variable.* Warszawa, PWN, 1968.

17. Kuczma, M.: *An Introduction to the Theory of Functional Equations and Inequalities (Cauchy's Equation and Jensen's Inequality).* Warszawa, PWN, 1985.

18. Kuczma, M.—Choczewski, B.—Ger, R.: *Iterative Functional Equations. Encyclopedia of Mathematics and its Applications.* Cambridge, Cambridge University Press, 1988.

19. Lanford, O. E. III: A computer-assisted proof of the Feigenbaum conjecture. *Bull. Amer. Math. Soc.*, **6**, 1982, pp. 427 – 434.

20. Li, T. Y.—Yorke, J. A.: Period three implies chaos. *Amer. Math. Monthly*, **82**, 1975, pp. 985 – 992.

20a. McCarthy, P. J.: The general exact bijective continuous solution of Feigenbaum's functional equation. *Commun. Math. Phys.*, **91**, 1983, pp. 431 – 443.

21. Marotto, F.: The dynamics of a discrete population model with threshold. *Math. Biosciences*, **58**, 1982, pp. 123—128.

22. May, R. M.: Simple mathematical models with very complicated dynamics. *Nature*, **261**, 1976, pp. 459 – 467.

23. Misiuriewicz, M.: Chaos almost everywhere, *Iteration Theory and its Functional Equations. Lecture Notes in Mathematics*, **1163**. Berlin, Springer, 1985, pp. 125 – 130.

24. Misiuriewicz, M.—Smítal, J.: Smooth chaotic mappings with zero topological entropy. *Ergodic Theory and Dynamic Systems* to be published (on appear).

25. Pelikán, J.—Gaisler, J.—Rödl, P.: *Our Mammals* (Czech). Praha, Academia, 1979.

26. Polyektov, P. A.—Pych, J. A.—Švytov, I. A.: *Dynamical Models of Ecological Systems* (Russian). Leningrad, Gidrometeoizdat, 1980.

27. Sierpiński, W.: *Cardinal and Ordinal Numbers.*Warszawa, PWN, 1958.

28. Smítal, J.: On boundedness and discontinuity of additive functions. *Fundamenta Math.*, **76**, 1972, pp. 245 – 253.

29. Smítal, J.: A necessary and sufficient condition for continuity of

additive functions. *Czech. Math. J.*, **26** (101), 1976, pp. 171 – 173.
30. Smítal, J.: A chaotic function with some extremal properties. *Proc. Amer. Math. Soc.*, **87**, 1983, pp. 54 – 56.
31. Smítal, J.: Chaotic functions with zero topological entropy. *Trans. Amer. Math. Soc.*, **297**, 1986, pp. 269 – 282.
32. Smítal, J.—Neubrunnová, K.: Stability of typical continuous functions with respect to some properties of their iterates. *Proc. Amer. Math. Soc.*, **90**, 1984, pp. 321—324.
33. Smítal, J.—Smítalová, K.: Structural stability of nonchaotic difference equations. *J. Math. Anal. Appl.*, **90**, 1982, pp. 1 – 11. Erratum, *J. Math. Anal. Appl.*, **101**, 1984, p. 324.
34. Šarkovskii, A. N.: Coexistence of cycles of a continuous map of the line into itself (Russian). *Ukr. Mat. Ž.*, **16**, 1964, (1), pp. 61 – 71.
35. Šarkovskii, A. N.: On cycles and the structure of a continuous mapping (Russian). *Ukr. Mat. Ž.*, **17**, 1965 (3), pp. 104 – 111.
36. Šarkovskii, A. N.—Peljuch, G. P.: *Introduction to the Theory of Functional Equations* (Russian). Kiev, Naukova Dumka, 1974.
37. Štefan, P.: A theorem of Šarkovskii on the existence of periodic orbits of continuous endomorphisms of the real line. *Commun. Math. Phys.*, **54**, 1977, pp. 237 – 248.
38. Targonski, G.: *Topics in Iteration Theory*, Göttingen – Zürich, Vandhoeck – Ruprecht, 1981.
39. Verejkina, M. B.—Šarkovskii, A. N.: Returning in one-dimensional dynamical systems. *Approximate and Qualitative Methods in the Theory of Differential-Functional Equations* (Russian). Kiev, AN USSR, 1983, pp. 35 – 46.

INDEX

Milton Keynes UK
Ingram Content Group UK Ltd.
UKHW040711141024
449569UK00005B/92